Innovation Design

Elke den Ouden

Innovation Design

Creating Value for People, Organizations and Society

Springer

Prof. Dr.ir. Elke den Ouden
Faculty of Industrial Design
Eindhoven University of Technology
Building HG3.34, Den Dolech 2
5612 AZ Eindhoven
The Netherlands
e-mail: E.d.Ouden@tue.nl

ISBN 978-1-4471-2267-8 e-ISBN 978-1-4471-2268-5
DOI 10.1007/978-1-4471-2268-5
Springer London Dordrecht Heidelberg New York

British Library Cataloguing in Publication Data
A catalogue record for this book is available from the British Library

Library of Congress Control Number: 2011938010

Cover design: eStudio Calamar S.L.

Printed on acid-free paper

Springer is part of Springer Science+Business Media (www.springer.com)

Preface

In October 2008 during the Dutch Design Week, an annual event in the city of Eindhoven, I held my inaugural lecture with the title of 'Designing (added) value'. While preparing the lecture, it became clear that the term 'value' was widely used but barely understood. It has different meanings in different contexts, and different disciplines use different definitions. So how can we ever create the 'shared value' for people, organizations and society that we aspire to so much? Growing numbers of people recognize that we are in a transition, in which 'business as usual' is no longer possible. The industrial paradigm focusing on shareholder value and profit maximization is being replaced by a new, transformational paradigm. One in which people are looking for ways to make meaningful contributions to addressing collective issues in which society is not just a stakeholder, but a beneficiary. People also expect this kind of ethical behavior from others and from organizations, both profit and nonprofit. At the same time there is a growing awareness of the tremendous business opportunities in 'doing good'. Social responsibility is no longer only about charity and philanthropy, but about creating value for people, organizations and society at the same time. Discussions with people in the industry and academia have clearly shown me that there is widespread recognition of the need for a better understanding of value, and of how to use it to create more meaningful innovations. Fortunately, my managers at Philips Innovation Services were among those who recognized this need. That gave me the opportunity to take a nine-month sabbatical to explore the subject more deeply and to write a book about it.

This book is aimed at practitioners in the field who are actively involved in both profit and nonprofit organizations: innovation managers, product managers, program managers, design managers, project leaders, architects, designers and engineers. I hope it will give them an insight into what meaningful innovations are, what value is, and practical methods and tools that can be applied directly in their own organizations. The scientific foundation also makes it suitable for (academic) educational purposes: academics at BSc, MSc and PhD levels studying innovation, design, business and/or engineering-related subjects, and in particular those aiming for a professional career in innovation.

Roughly the first half of the book explores 'value'—its meaning from different perspectives and in different contexts. This exploration is based on many years of discussions in projects in the industry and academia, as well as an extensive literature study conducted more recently. It considers value from the viewpoints of users, organizations (profit and non-profit) and society, as well as from the perspectives of economics, psychology, sociology and ecology. The resulting Value Framework can serve as a map to create meaningful innovations that improve the quality of life of users, provide value for organizations, enable ecosystems to evolve and improve the quality of life for society as a whole. In other words, caring for both people and planet. The second half discusses how to get started on creating meaningful innovations, and provides the practical tools needed to really create the value that is aimed at for the various stakeholders.

It became clear at an early stage that creating meaningful innovations with value for organizations, users and society at large most often requires collaboration between different players. In fact it is not just the innovation itself that needs to be designed, but also the ecosystem to support it. These multiple players are needed to jointly contribute knowledge, experience and resources; to deeply understand the societal issue being addressed; and to generate ideas that really solve that issue in a way that brings value to all the stakeholders. An issue like obesity in children cannot be solved by a single organization. It needs the collaboration of parents, schools, the food industry, governments and many other people and organizations that, in one way or another, are all part of the system influencing children's eating habits and activity levels. Failure to recognize this fact could lead to 'innovations' and other measures that would be unlikely to achieve the desired goal of happy and healthy children. Moreover, in many cases the solution is not just a single product or service but a system of combined products and services from different organizations. This requires a sound business model to ensure a sustainable overall value proposition, as well as sustainable value for all the social and economic actors in the ecosystem. It may sound complex and challenging to address societal issues at this level, in creating innovations that really do bring value to multiple stakeholders. But I hope this book will provide the necessary insight and practical tools to get started. It gives many inspiring examples to help readers take up the challenge and contribute meaningfully to solving the societal and economic challenges we are currently facing – challenges like the aging population, energy scarcity, environmental issues, social well-being and many more.

Chapter 1, 'The Age of Transformation', provides the context of the book. The world is currently going through a transformation. We face a number of tremendous societal challenges for which traditional approaches no longer work. Mankind has gone through big changes before: moving from the agricultural economy to the industrial economy, and more recently into the 'experience' economy and the knowledge economy. This chapter explores past and present paradigm shifts to set the scene for what follows. Most of all, facing the current societal challenges will require collaboration between profit and nonprofit sectors as well as connected individuals, to create meaningful innovations that will guide people and society through the transformation.

Chapter 2, 'Levels of Value', explains that the transformational solutions needed to tackle today's societal challenges must create value not only for their users and the organizations that market them, but also for society as a whole. The term value is used frequently, but has different meanings in different contexts. This chapter explores the different levels of value that are relevant for innovations addressing societal challenges: the user, the organization, the ecosystem and society.

Chapter 3, 'Value from Different Perspectives', takes a deeper look at the meaning of value. Value is a subject that has been studied in many contexts and many scientific disciplines. Even within one discipline it has different connotations. This chapter gives an overview of 'value' from the perspectives of the social sciences that provide a relevant context for innovation: economics, psychology, sociology and ecology. It also looks into the history and etymology of the term value, and the philosophical perspective, to provide a sound basis for further discussion throughout the book. The different views on value are integrated into one overall Value Framework. This combines the value concepts from economics, psychology, sociology and ecology, and explains them at the four levels of the previous chapter: user, organization, ecosystem and society. The Value Framework serves as a basis for design to create shared value with transformational innovations.

Chapter 4, 'Meaningful Innovations', builds on the Value Framework to show how to define meaningful innovations. Finding solutions for today's societal challenges demands new value propositions for people, organizations and society at large. Transformational innovations will only have an impact on society if they bring about changes of behavior and create long-term value for users. Moreover such innovations are likely to disrupt the current industry structures and ecosystems. The Value Framework shows how to define meaningful innovations that combine creating an 'experience' for the user, 'doing well' for the organization, 'doing good' for the total ecosystem and creating a 'transformation' at societal level.

Designing meaningful innovations that create value for users, organizations, ecosystems and society requires a holistic view to integrate seemingly conflicting needs and requirements into compelling solutions. Chapter 5, 'Creating Meaningful Value Propositions', describes and shows through examples the way of thinking, the process and the tools to reach integrative value propositions and their accompanying business or value models. It shows that innovation projects of this kind typically have higher uncertainties. These translate into a need to integrate knowledge and experience from various disciplines, and to follow a more explorative approach in the initial phases of the projects. Building on the literature from design, marketing, business and innovation management, as well as extensive practical experience in projects in the field, a practical approach is defined that describes pragmatically how to define and realize meaningful innovations. It contains descriptions of methods and ways of working, supported by examples.

The final chapter, Chap. 6, 'Designing Ecosystems', shows how new ecosystems can be designed for solutions that require combinations of products and services from different organizations. This will often be the case in defining meaningful innovations that provide sustainable value for people and society. A number of organizations are needed to realize the innovation, hence apart from the products and services themselves, the total ecosystem needs to be designed. This has to include all the relevant social and economic actors required for a successful launch of the innovation on the market, but also for sustained service in the long run. The design of the ecosystem needs to ensure a return on investment of both tangible and intangible value for all the business parties and other stakeholders involved. This chapter shows how new ecosystems can be designed by presenting the Value Flow Model as a method to identify the relevant stakeholders and the values that are important to each of them, and to balance those values in the total system. The method has proved to be valuable in enriching value propositions, but also in gaining commitment from the different business actors to make the investments required for implementation. The visualization of tangible and intangible value flows enables the balancing of value across the different parties to ensure sustainable value for all.

This book aims to provide a solid background on value creation and the process of defining new value propositions, business models and ecosystems for meaningful innovations. It builds on practical experience and an extensive literature overview. Although I believe that the literature included is representative of the different fields, I do recognize that I am not an expert in all the fields mentioned. So I would like to apply a 'probe and learn' approach to this book. I also invite readers to contribute to the second edition by pointing out missing but relevant topics and literature. If you would like to share your ideas on implementing my thoughts in practical projects, you are very welcome to do so. I will keep a list of suggestions and examples of implementations online at: www.elkedenouden.nl/innovationdesign.

I hope this book will inspire readers to embark on the journey of designing value propositions that bring value to people, organizations and society. And that, through them and their ideas, it will contribute to making the world a better place!

May 2011 Elke den Ouden

Acknowledgments

A book like this is the result of the thoughts, experience and efforts of many. My colleagues at Philips and the university and the many people I have met over twenty years of working on innovation projects have all contributed to the growing insights that provided the background for this book. But I would not have started the process without the encouragement of Allard to follow my heart and take the time to dive deeply into the subject. I am grateful to my managers at Philips Innovation Services, Iason Onassis and Ruud van Vessem, to give me confidence and allow me a sabbatical for this purpose.

I would like especially to thank Rianne Valkenburg for the discussions that we had at moments when I lost confidence in my ability to complete the job. After I had read over 600 books and articles on value from different perspectives, and sometimes couldn't see the wood for the trees, she helped me tremendously in creating the first rough version of what was to become the Value Framework. I am also grateful for her feedback on the draft text, and for the in-depth discussions on what makes transformational design different from more traditional design processes. I would also like to thank my colleague Dirk Snelders for his valuable feedback on the draft text from a scientific and educational perspective. And my friend Pauline van Hezik, who has been an inspiration over the years in looking at deeper levels of being more mindful.

Nigel Morrison has been a great help in improving the readability of the text, and I appreciate his patience in cutting my long sentences into bite-size pieces. I would also like to thank my sister Sofie, for her help with the graphics, and especially the cover design. Anthony Doyle and Claire Protherough of Springer deserve special thanks for guiding me through the publishing process, which was new to me.

Grateful acknowledgement is made to Philips and the Design Initiative for the permission to use their material. I would also like to thank the individuals and publishers that gave me permission to quote or use illustrations from their work.

I owe a lot to my close friends and family for their support and understanding during the months in which the book affected their lives as well. Thank you for all your support in the past, for the great things that we have done together, and for being there.

Last but not least, I would like to thank all those people who are working on meaningful innovations, and by doing so are creating more value for society. You make writing a book like this a worthwhile endeavor.

Contents

1 The Age of Transformation . . . 1
Introduction . . . 1
Threats... . . . 2
...Are Opportunities. . . . 3
Paradigm Shift . . . 5
Towards the Transformation Age . . . 6
References . . . 10

2 Levels of Value . . . 13
Introduction . . . 13
Value for the User . . . 13
Value for the Organization . . . 15
Value for the Ecosystem . . . 17
Value for Society . . . 18
References . . . 19

3 Value from Different Perspectives . . . 21
Etymology . . . 21
The Philosophical View . . . 22
Integrated View . . . 23
The Economic View . . . 23
Economic Value Concepts . . . 29
The Psychological View . . . 30
Psychological Value Concepts . . . 34
The Sociological View . . . 37
Sociological Value Concepts . . . 42
The Ecological View . . . 46
Ecological Value Concepts . . . 51
The Value Framework . . . 55
References . . . 56

4 Meaningful Innovation 61
Introduction 61
Defining New Value Propositions 62
Value Propositions for the User. 66
Experience 73
Value Propositions for the Organization 75
Doing Well 81
Value Propositions for the Ecosystem 82
Doing Good 85
Value Propositions for Society 87
Transformation 90
Meaningful Innovation 92
References 93

5 Creating Meaningful Value Propositions 97
Introduction 97
Innovation Approach 99
Inspiration 108
Ideation 113
Implementation 122
Designing Business Models for Transformational Innovations 127
References 138

6 Designing Ecosystems 143
Introduction 143
Defining Ecosystems 145
Innovation in Ecosystems 149
Designing New Ecosystems 153
The Value Flow Model 154
Select Parties for Ideation and Enrich the Value Proposition 162
Identify Stakeholders and Their Interests 164
Define Roles and Value Flows 171
Selecting Parties for Realization and Implementation 181
Evolution of Ecosystems 185
References 188

Index 193

Chapter 1
The Age of Transformation

Introduction

Whether you are a pessimist or an optimist, there is one thing we all agree on: things are changing. The world as we know it would not be the same in the near future. A transformation is taking place, and it is going to affect all of us.

In the pessimistic view, it is inevitable that the world as we used to know it no longer exists. Mankind has created a mess in the ecological system. We are squandering our planet's resources and violating nature, which has provided us with everything we need for so long. The pursuit of wealth has resulted in an increasing gap between the rich few and the poor masses. Our privacy is at risk with the widespread use of social media and video cameras on every street corner. Unhealthy lifestyles have caused an epidemic of obesity, and despite the advances in medicine and technology the elderly still are not living happier lives than before. The habits that we developed long ago of gathering food and the necessary supplies to exist have developed into just gathering stuff we rarely use. We have lost touch with nature, so we do not have a clue about what it takes to produce the stuff we buy and consume. We are using up natural resources at a speed that is completely out of tune with the rate at which they were formed. So the pessimists are warning us: we are racing full speed ahead on a dead-end track. So we had better brace ourselves for the crash that is sure to happen sometime soon.

Optimists see things differently. The ongoing developments in technology provide us with ever more opportunities to face the challenges ahead. No problem is so big that we cannot solve it with all the ingenuity we have available . A crisis is something positive, because it drives us to use our creativity to find new solutions. The information age has brought rich and poor together: the world has become flat. People in the 'western world' are no longer tied to fixed jobs for life in corporate or governmental organizations. They take the opportunities of self-employment, and follow their dream of living a life in which they build on their strengths and have more freedom to choose what they do. Recent developments in the Middle East show that undemocratic regimes that have ruled for decades can

E. den Ouden, *Innovation Design*, DOI: 10.1007/978-1-4471-2268-5_1,

suddenly be overthrown by masses of people who—through social media—understand that they are not alone. People in developing countries are no longer stuck in poverty forever. All those with drive and a bit of money—whether it is their own or a microcredit loan—can make themselves useful and contribute to the world's well-being. So optimists are encouraging us to prepare for self-management and empowerment, and to start contributing to and benefiting from the plentiful opportunities that lie ahead.

Both these views are valid, and they are not mutually exclusive: we do not have to choose because...

Threats...

Plenty has been written on the multiple crises we are currently experiencing. The economic crisis has not gone away. Experts may differ in how they describe it—a double dip or an L-shaped recession—but what they do agree is that it will take a while before the economic markets are back to where they were before the bank crashes. If they ever do that...

On the health front, we are confronted with viruses and bacteria that are mutating and threatening to infect large numbers of people. We are living so closely together that a virus could easily create an epidemic. Newspaper headlines talk about a pandemic each time a new virus is discovered. In recent years there have been SARS, bird flu and swine flu, making it clear that we are very vulnerable as a species. We are threatened by things we can hardly control, but even where we can control our health we do not necessarily seem to care too much. Many people enjoy unhealthy lifestyles, with unbalanced diets causing malnutrition, and not having enough physical activity. All this results in too many people being overweight or even obese, with the associated health effects.

Our population is ageing. Of course it is a good thing that life expectations are much longer, thanks to better standards of living. Today, many conditions that were fatal in the past can be cured or kept under control. Patients can live longer, with a higher quality of life. But this is also a threat to the costs of the health care system. Some predict that in a few years we will need almost half our workforce to work in health care if we want to maintain the present care levels. Apart from the fact that we would not be able to find so many trained staff at such short notice, you do not need to be a wizard in business economics to calculate that we will never be able to afford the associated costs. Solutions that aim to increase the efficiency of health care actually reduce the amount of social contact between the patient and the caretakers, which then increases the feeling of loneliness. Lonely patients could feel more depressed and anxious about their health. This might in turn lead to more frequent visits to the doctor and regular use of medicines, thus in fact to higher costs of health care.

Whether mankind is the cause of climate change or whether it is caused by natural phenomena is the subject of ongoing debates. But whatever the cause, there are certainly huge disasters taking place in areas that looked so safe before. Floods, earthquakes, bush fires, tsunamis and volcanic eruptions are claiming many lives. Whole areas are affected, and many people have lost everything they had.

Some natural disasters are certainly out of our control, but that does not mean we are not responsible for squandering natural resources at a tremendous rate. While some cultures and people believe nature is sacred, they have long lost the power to protect it. Virgin forests are sacrificed for economic gain and to plant crops for biofuels. Not only are rain forests important for our oxygen, but nature in general also has a very positive impact on our general well-being. So the disrespect we are showing might not turn out to be such a good idea.

Wherever there is gas, oil, gold, diamonds, gallium, uranium or whatever resource to be found, we swoop on it and extract it. And although we are becoming more efficient by the day, prices of commodities are going up as demand soars. The list of minerals that will be scarce in the foreseeable future is constantly increasing. This trend will certainly affect the availability of materials for all kinds of products.

Social cohesion is also under threat. In the pursuit of wealth, people at all levels take decisions that seem to mainly address their own interests. People who are only looking for the fastest way to fill their own pockets are managing institutions with the responsibility of caring for the weakest members of society. When management of institutions and companies alike show this kind of behavior, we cannot expect employees to keep working for the greater good of society. It is almost as if you are a complete dummy, just working hard and not taking any short cuts to achieve your personal goals ever faster. As a result social fabric is weakening, with fewer people feeling inclined to take care of each other.

... Are Opportunities

So we can choose to be paralyzed in despair, or we can choose to see the abundance of opportunities that these multiple crises bring. Many innovative solutions have been found in times of crisis. There is a growing awareness that things need to change. More and more people are willing to invest time and resources in making a change. And as many examples have proven, the problem is not the lack of financial resources but rather a lack of creativity to think differently, and to find new solutions that may start small but can grow into a sustainable ecosystem that develop further over time.

Just one example: Muhammad Yunus visited the poorest households in the village of Jobra, Bangladesh, in 1976, where women made bamboo furniture. These women had to take loans from moneylenders at extreme rates to buy the bamboo, leaving them barely any profit to live on. So he decided to lend US$ 27 out of his own pocket to 42 women. Subsequently he was able to get a government loan to extend his clientele, and in 1983 he founded Grameen Bank. Today the bank has lent more than $8 billion to 8.1 million borrowers, 97% of them are women, and changed the lives of 80% of the Bangladeshi families. This example shows that starting small can create a big impact.

What is different in tackling societal challenges, compared with traditional engineering, design or business challenges is that they need a more holistic view of how the solution needs to be defined. A traditional engineering or design problem can be solved at the level of a product. Although engineering or design problems can be challenging because of contradictory requirements, the solution can usually be found within the scope of the product itself, and so the decision can be taken by the lead engineer or product manager. A business problem at the level of a company and its market can usually be solved within the scope of the decisions that a manager can take. But finding real solutions to societal challenges requires the consideration of all these aspects at the same time; it crosses the borders of a single decision-maker. It requires knowledge at the level of the user of the prospective products and services, the level of the organization that will bring those products and/or services to the market, the level of the ecosystem that links the various products and services to their users and other stakeholders, and the level of society that will reap the benefits of the solution. By linking the views, opportunities and resources of these levels, the goal becomes clearer and solutions can be strengthened.

This may look like an insoluble problem, but in fact societal challenges provide tremendous opportunities for business [11]. Social innovations are motivated by the goal of meeting a social need. They aim to fill the gap between what there ought to be, between what people need and what they are currently offered by governments, private firms or non-governmental organizations [4]. Social innovations bring new ideas that improve quality of life. This concept is different from business innovations, which aim to create a profit for a specific company. Yet the concepts are overlapping and social innovation provides opportunities for business to go hand in hand with improving quality of life [10]. This business potential is real—not just an opportunity to do a charitable favor to society, as is unfortunately too often the current practice in many social responsibility programs in companies [11].

Solving societal challenges requires a good understanding of what prospective users are faced with. For example, a good understanding of the experience of elderly people who regularly take medicines and visit their physicians could reveal that at a deeper level they feel lonely and anxious about their health. Alternative solutions to medication and physician visits could then include addressing their loneliness. Solutions at this level would not only increase the quality of life of the patients, but would also decrease the healthcare costs.

Innovation can also start at the level of the company. Continuous development of technologies makes new solutions feasible that were previously beyond imagination. Business opportunities can be found in the links between the emerging technological solutions and the challenges in society. We can also start at the level of societal challenges, with brainstorming, for example, on solutions to improve quality of life or ways to reduce the ecological footprint of citizens in developed countries.

Good solutions are those that create value both at user level and for the organizations providing them, without harming the planet. But we need more than just good solutions.

Paradigm Shift

What is needed is aiming higher than just providing good, although partial, solutions. Partial solutions will never radically reduce the energy consumption of the world's citizens, while radically improving quality of life at the same time. We need more than partial solutions to achieve that.

The conventional business perspective shows that we are fine if we address a big enough target customer group to buy a solution at a price that gives us a good margin over the development and production costs. But to make an impact at societal level, we need as many users as possible to adopt the solution and keep on using it. This means two things: we need to make a compelling offer that users are willing and able to buy, and moreover we also need them to use it extensively. For example, to tackle obesity among children we could launch a game that needs them to be physically active. If we take a narrow view, we can claim the product is a success if more than the expected numbers are sold at the recommended retail price. But in a wider view, it can only be a success if the children are indeed more physically active over a longer period of time and eventually loose weight. So it is not enough just to sell those products, they also need extended usage to provide longer-term societal benefit. And to do that, the game—in this example—has to create sustainable behavioral change in the children using it.

The societal challenges that we currently face are not isolated problems that can be solved by single solutions sold just like 'normal' products. We need to change the behavior of as many people as possible to make an impact. Energy scarcity cannot be solved just by a few people who drastically change their behavior, and nor can it be solved by everyone doing just a little. As David McKay put it: "if everyone does a little, we'll achieve only a little" [3].

To change the behavior of people on a larger scale, we need to provide solutions that they love to use. Using the product or service should provide a positive experience; one that persuades them to use it more often, for a longer period of time. It also means that the organizations involved in providing those products and services need to benefit from the solution. The business case should be sound, so that they get a return on investment within a reasonable timeframe. But the business model should also allow them a sound business in the longer run.

The ultimate solution is therefore the one that provides a pleasurable experience for end-users, with the ability for the organizations involved to do well, sustainably. Only in this way will there be a big enough positive impact at societal level, and will real transformation take place.

Towards the Transformation Age

There are signs that the paradigm shift that is needed is already taking place. It is the next one in a series of paradigm shifts that have happened in the last few decades. From the agricultural economy we went into the industrial economy. From the industrial economy we went into the experience economy. From the experience economy we went into the knowledge economy. And now we are entering the transformation economy. To understand the impact of this, let us first look back at those earlier shifts.

The Agricultural Economy In the rural, agricultural economy people made a living by growing and selling crops and livestock at the market. As they were dealing with living things, they were largely dependent on the seasons and climate influences. Although some may reduce the task of agriculture to the production of food (as an industrial process), a wider view shows that it also contributes to other things [13]: it keeps man in touch with living nature, of which he is and remains a highly vulnerable part; it contributes to humanizing man's wider habitat; and it brings forth food and other materials that are needed.

If production of those goods is in line with consumer demand, farmers can have a good living. But this came to an end when the drive to increase productivity led to investments in machinery, and ultimately to overproduction. Prices dropped, and the necessary investments made the costs higher. It became harder to have a good living with farming. At the same time a shift to an urban, industrial economy was taking place, and for many farmers it looked more attractive to abandon their farms and move to the city in pursuit of jobs in industry.

The Industrial Economy The urban, industrial economy differs from the agricultural economy in many ways. The focus on production of non-perishable goods and commodities such as textiles, coal and metal meant larger volumes could be produced and kept in stock or be distributed over larger distances. Investments in technology such as steam power, together with the availability of cheap labor from farmers migrating to cities, made it possible to achieve economies of scale that allowed both greater quantities and higher quality at the same time. The revolution in transportation through the construction of canals and roads, as well as the railway system, increased the ease and speeds with which raw materials and finished goods could be transported to and from the urban centers.

The industrial revolution caused unprecedented changes in society, changing the whole way of life for many people in an unusually short period of time. Where earlier, extended families stayed in the same area for many generations, people suddenly became more mobile and bonds with the extended family became less strong. Workers sold their labor in return for a wage or salary, and were no longer free and self-sustaining. This lead to criticisms, as in some cases workers were exploited. The factory working environment was often harsh: noisy and dangerous, with mind-numbing repetitive tasks. Nevertheless, the industrial age saw a vastly improved standard of living for many people.

For companies, the increasing competition from global companies has caused tremendous pressure on the price and quality of their products. Availability of goods from countries with lower wages has made companies in the West shift their manufacturing operations out of the Western world. Looking for alternatives to build a sound business, many have shifted focus to create higher added value through services and experiences.

The Experience Economy The growth of services as a sector is a logical consequence of the decline of industrial activities. Services are considered as economic goods consisting of labor, advice, managerial skills etc. [7]. Service industries account for an ever-increasing proportion of the national income in nearly all developed countries. Services differ from the production of goods in a number of respects [5][14]: their intangibility; the fact that customers are directly involved and participate in the delivery, and therefore the production and consumption takes place simultaneously; their perishability; and the differences in experience of different consumers. Where competition is strong, and customers can choose from a wide variety of offers, it becomes important to distinguish the offering. The distinguishing factor in services is the experience. Buyers of experiences want to be engaged by what the company reveals over time. Companies strive to stage customer surprise, exploiting the difference between what the customer gets to perceive and what the customer expects to get [9]. This is quite a contrast with the 'industrial' notion of customer satisfaction that strives to meet the expectations of larger customer groups. In creating an experience a personal element is taken into account, and the aim is to surprise the customer. For returning customers, it also means that the company has to keep a record of the past experiences of each specific customer because just providing exactly the same as in the past might not create the desired surprising effect. To continue to appeal to customers, organizations need to carefully manage the experience they create.

The Knowledge Economy The knowledge economy is part of the service sector, but distinct enough to be classified separately: the quaternary sector focuses on the knowledge-based part of the economy. It includes economic activities concerned with information: its acquisition, manipulation and transmission [8]. The category also includes the fast growing economic activities in the field of media. The knowledge-based economy generates information, rather than goods and services, and successfully creates new knowledge as the basis for new products and services [8].

The rise of knowledge-based services is enabled by the widespread use of information technologies. Accurate, up-to-date information from professional services such as news agencies, weather stations, national statistics institutes etc. is put online. The amount of data that is available online is growing very rapidly, and it can be used to provide knowledge-based services or to enhance existing services with additional information. For example a travel agency used to provide general climate information on the destination, but now the e-ticket comes along with accurate forecasts of actual weather conditions at the site.

Another enabler for the knowledge-based economy is the widespread use of automated data generation. In personal navigation, for example, the logging of the speed of mobile phone users on highways allows accurate, up-to-date estimation of travel times, with real-time allowance for delays due to road works, accidents or traffic jams. This information, provided to users of TomTom's HD Traffic service, gives much more accurate indications of estimated travel times to specific destinations than average travel times.

A third factor that enables the knowledge-based economies is the ever-growing online community that contributes to building knowledge and making it available to others. A good example is the use of 'wikis': websites whose content can be changed and extended by users through their web browsers [5]. Wikis range from small ones for private, department or company use, right up to large, worldwide internet sites as Wikipedia, the popular internet encyclopedia. What makes these concepts different is that they are created by volunteers. The accumulation of small contributions by many individuals can create a tremendously rich wiki.

The knowledge economy also provides new opportunities for organizations. New ways of working together are enabled by technological advances, allowing radically superior forms of innovation. Cyberteams of self-motivated people with collective visions enabled by the web to collaborate in achieving common goals by sharing their ideas, information and work are innovating in collaboration. In these networks, knowledge workers collaborate and share in internal transparency. They communicate directly rather than through hierarchies. And they innovate and work toward common goals in self-organization instead of being ordered to do so. The World Wide Web itself and Linux are examples of successful innovations in such networks [2]. These examples also show that new business models came into use with the knowledge economy, and the traditional direct exchange model has been replaced by services that are offered for free, but have other ways to generate revenues such as through advertisements or 'clicks'.

We are still at the beginning of the knowledge economy, and there are plenty of opportunities to further build new knowledge-based services, to innovate through the use of knowledge networks, and to use supercomputers to make sense of the increasing amount of available data. But some effects of the knowledge economy are already visible. Knowledge workers are much less dependent on their employers than they used to be in the industrial age. The investment needed to set up a business in the knowledge economy is extremely low: a PC with an Internet connection will do. For people who want more control over their daily activities, it is economically feasible to become self-employed and pick the projects they want to do. They benefit from more freedom in what they do, and when they do it. Knowledge workers are not bound by regular office hours. For companies this means additional flexibility. But there is also a challenge: they will need to stay attractive to the more flexible workforce, as good people are much more mobile than before.

The knowledge-based economy initiated the emergence of a fifth economy: the transformation economy.

The Transformation Economy Although not officially announced as such, there are signs that the next important economy will be the transformation economy. For two reasons: it needs to be, and it can be. Transformation is defined as a marked change in form, nature or appearance [6]. A European Union think tank expects that the next 20 years will be disruptive: a new multi-polar world is emerging in which power is more diffuse and international dynamics is more complex, against the background of sweeping societal transformations that include an ageing society, declining domestic supplies of labor and technological and communications revolutions that are changing both private and professional relationships and imposing new patterns of life [12]. In this age, transformation will take place at societal level, through large numbers of individuals willingly contributing to it. This transformation is needed because of the abundant negative side effects of the industrial age: the threats referred to before are real. This transformation is also possible because by building on the knowledge-based economy, and on the willingness of many people to contribute to higher goals and live more meaningful lives, we can seize the opportunities.

To a certain extent, we need to re-establish the virtues of the agricultural age: reconciliation of man with the natural world. And as Schumacher already noted in 1973: this cannot be achieved by tourism, sightseeing or other leisure activities. We will need to focus all our actions to the three-fold ideals of health, beauty and permanence [13].

Pine and Gilmore define transformations as have lasting consequences beyond their consumption. Buyers of transformations seek to be guided toward some specific aim or purpose, and transformations must elicit that intended effect. A transformation requires a change in attitude, performance, characteristics or some other fundamental dimension; it affects the very being of the buyer, in fact the customer *is* the product. The change has to be permanent; if it is only temporary and not sustained, then it was not really a transformation but merely a momentary uptick along the same old journey [9].

This description provides a good insight into what we need to achieve. Typically the aspirations that we need to address should contribute to society as a whole. They include aspirations like changing our energy consumption, reducing our ecological footprint or living a healthier lifestyle, with a more balanced diet and more exercise. We will need to understand the personal differences between individual people, and to address these in the offering. 'One size fits all' is a concept that already became outdated in the experience age, and it is even more important to make sure that individual differences are addressed if we want to create sustained behavioral change.

The transformations we are aiming for will have to provide interesting and meaningful challenges for the collaborative innovation networks that generate ideas, as well as sound business for the organizations that provide the solutions. In fact transforming propositions need to provide value at different levels: for the user, for the organization and for society at large. These different levels of value are explored in more detail in the next chapter.

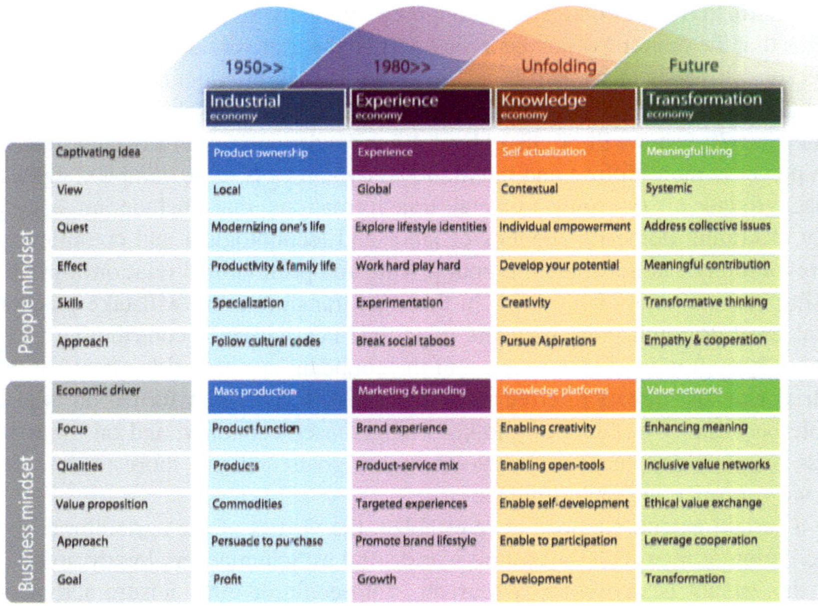

Fig. 1.1 Paradigms in value creation: extension of the model of Pine & Gilmore [9] by philips design [1]

Figure 1.1 shows an overview of the most important distinctions between the economies described in this chapter. The table is an extended version of the one provided by Pine and Gilmore [9]. For each of the economies, the figure shows how value is created from a people as well as a business perspective.

The answer to the current societal problems is not another economic reform. On the contrary, a dynamic equilibrium between the economic, social and environmental dimensions needs to be restored [12].

References

1. Brand, Reon, and Simona Rocchi. 2011. Rethinking value in a changing landscape. A model for strategic reflection and business transformation. A philips design paper.
2. Gloor, Peter A. 2006. *Swarm creativity: Competitive advantage through collaborative innovation networks*. New York: Oxford University Press.
3. MacKay, David J.C. 2009. *Sustainable energy—without the hot air*. Cambridge, UK: UIT Cambridge Ltd.
4. Mulgan, Geoff. 2007. *Social innovation what it is, why it matters, and how it can be accelerated. Paper from the young foundation*. London: The Basingstoke Press.
5. Oxford dictionary of business and management. ed. Jonathan Law, Oxford University Press 2009. Oxford Reference Online. Accessed 28 December 2010.

6. Oxford Dictionary of English (revised edition). ed. Catherine Soanes and Angus Stevenson. Oxford University Press, 2005. Oxford Reference Online. Accessed 7 December 2010.
7. Oxford Dictionary of Finance and Banking. ed. Jonathan Law and John Smullen. Oxford University Press 2008. Oxford Reference Online. Accessed 28 December 2010.
8. Oxford Dictionary of Geography. Susan Mayhew. Oxford University Press 2009. Oxford Reference Online. Accessed 28 December 2010.
9. Pine, B.Joseph, and James H. Gilmore. 1999. *The experience economy work is theatre and every business a stage*. Boston, Massachusetts: Harvard Business Press.
10. Pol, Eduardo, and Simon Ville. 2009. Social innovation: Buzz word or enduring term? *Journal of Socio-Economics* 38: 878–885.
11. Porter, Michael E., and Mark R. Kramer. Creating shared value. Harvard business review, January 2011.
12. Reflection Group on the Future of the EU 2030. Project Europe 2030. Challenges and Opportunities. A Report to the European Council, May 2010.
13. Schumacher, E.F. 1973. *Small is beautiful: A study of economics as if people mattered*. London: Blond & Briggs Ltd.
14. Zeithaml, Valarie A., A. Parasuraman, and Leonard L. Berry. 1985. Problems and strategies in services marketing. *The Journal of Marketing* 49(2): 33–46. Spring.

Chapter 2
Levels of Value

Introduction

The previous chapter explained the need for transformational innovations to provide solutions to the challenges facing our society. Awareness is growing that these societal and environmental challenges need a collaborative effort by large corporations, innovative entrepreneurs, non-profit organizations, governments and citizens. If they are well designed, transformational and innovative solutions create value for their users and for the organizations involved in developing and delivering them, and as a result they generate sustainable value for society at large.

But what does value actually mean in this context? The word value is used very often, but it certainly does not mean the same to everyone in every context. To better understand value, it is important to distinguish the level at which it is perceived. Four distinct levels of value are relevant in the context of innovation: value for users, value for organizations, value for ecosystems and value for society. As shown in Fig. 2.1, these levels overlap. The values of the user need to be included in those of the organization, those of the organization need to be included in the values of the ecosystem and those of the ecosystem in turn need to be aligned with those of society. Transformational innovation addresses value for all four levels at the same time. But value has a different meaning at these different levels, and it is important to have a good understanding of what these levels are, and what value means at these different levels.

Although the concept of the levels is more general, and could be applied for various purposes, the discussion of the levels in this chapter is limited to the context of innovations that aim to create value through transformation.

Value for the User

The first level addresses value for the user. The user is the ultimate target of innovation. He is the client, who is expected to use the system, product or service. For transformational innovations it is important that the use of the product is a

E. den Ouden, *Innovation Design*, DOI: 10.1007/978-1-4471-2268-5_2,

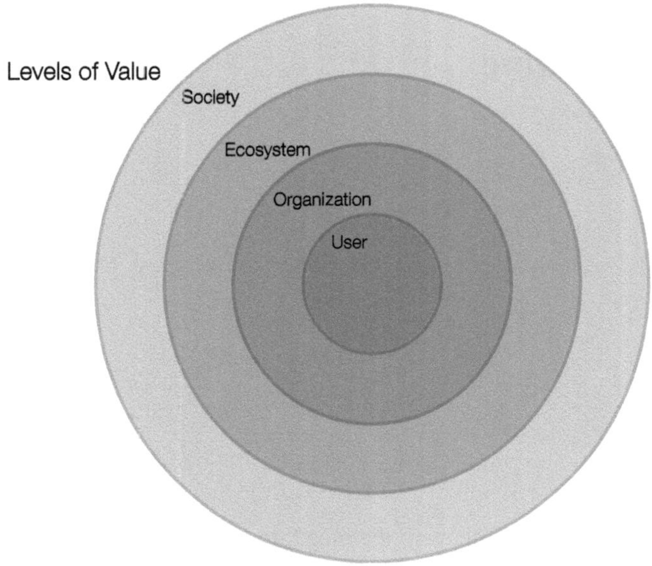

Fig. 2.1 Levels of value

pleasurable experience. If the user does not continue to use the product for an extended period of time, it cannot be claimed to be a success. This differs from the traditional business view which focuses more on the sale of the product, and is less concerned with its extended use over time.

The buyer and the user may be the same person, but often they are not. The person buying the product may buy it for someone else as a gift, or on behalf of someone else. The buyer may be an individual, such as a family member, but may also be someone from an organization. In healthcare, for example, doctors or insurance companies in many cases take decisions for patients. This wider notion of people involved in the total process around decisions and the implementation of innovations will be addressed in the section on ecosystems.

Value for the user addresses the definition of a value proposition that is attractive to the user. That means it is important to deeply understand the motivational values of the user, especially for innovations that aim to change user behavior [6]. If these values are not taken into account users may stop using the product or service, and the potential value of the innovation at other levels is then lost.

If a larger user group is needed to create an impact at societal level, it is also important to understand the differences between individual users and user groups. Both cultural differences [3] and personal preferences may influence the adoption of a product or service [20]. Users' motivations may differ. One person may be prepared to sacrifice some comfort to achieve a particular goal. For example some people do not mind turning their home thermostat down a few degrees to save energy, and just wear warmer clothes. Others may not be prepared to make such an adjustment. It is therefore risky to judge a larger group of people on the perception of a few representatives.

The ultimate solution for transformational changes is not to require users to make sacrifices. Instead the aim should be to create such a pleasurable experience that they will use the product even more often than they first envisioned. A transformational innovation helps the users to look differently at themselves and change their actions and behavior. Moreover, if the experience is pleasurable it will also help the widespread adoption of the innovation, as users will recommend the product or service to others [19].

Value for the Organization

The second level of value concerns the organization. Organizations are required to design, produce and market new, innovative products and services. An organization in this context can be a for-profit or a non-profit organization that is innovating to create sustainable value for itself by providing added value for its customers. At the same time, it also creates value for their employees by providing jobs.

In business management, an organization is characterized by a more formal pattern of social relationships, and by having a defined goal. Often the overriding purpose and objectives are encapsulated in a mission statement, which is used to communicate this purpose to all stakeholder groups, both internal and external, and to guide employees in their contribution towards achieving it. The strategic intent reflects the direction in which the organization intends to develop. It should be both understood and interpreted by all employees in relation to their work, and is a crucial element in strategic management [15]. Many business managers regard the process of creating value largely as a top-down endeavor. Innovation is regarded as one of the business processes of an organization, and mostly these processes are guided by targets from the top. The organization's vision of an attractive market or competitive opportunity is implemented into innovation projects. These follow an efficient innovation process to translate the vision into a market reality [5].

Organizations are also social systems that interact with other social systems whose values and goals are oriented towards those of the wider society. Key requirements for organizational maintenance are those that apply to all social systems: adaptation, goal attainment, integration and value maintenance. Sociologists cast doubt on whether it makes sense to refer to organizations as institutions, which pursue organizational goals. There are many studies that show that official goals may bear no relationship with actual or operative goals; that organizations frequently have multiple and conflicting goals; and that goal displacement may occur [16]. The informal culture of work within organizations has a significant impact on what is done and how it is done. Many good ideas for innovative solutions to meet market needs or improve performance arise spontaneously from staff working either directly or indirectly with customers. For this bottom-up innovation to flourish, the organization must encourage the development of organizational creativity at all levels, a key element of any innovation culture [5].

Organizations may choose between different strategies to create value. The classic strategies for competitive advantage as defined by Porter are: lowest cost, differentiation and focus [18]. These can be translated into three value disciplines: operational excellence—offering the best price and/or the least inconvenience to the customer; customer intimacy—understanding the fulfillment of customer needs; and product leadership—delivering unique value products [21]. But as Ohmae already emphasized more than 20 years ago: a value-adding strategy is created not by setting out to beat the competition but by setting out to understand how best to provide value for customers [14]. This also requires a shift from a focus on the organization and its needs to a focus on the end-user, so that the assets that represent real value can be identified [23]. Innovation requires constant redefinition of value by correctly guessing the customer's unarticulated desires and applying the organization's expertise to satisfy them [5]. This means that it is not enough just to listen to the voice of the customer. This provides valuable input on the user, but will not drive new innovations. Transformational innovations by definition address unmet needs, which users are often unaware of. This means organizations need to come up with more radical solutions that create new markets. Or as Verganti stressed: traditional market-pull methods of innovation—which scrutinize customer acceptance before releasing products to the market—sometimes even restrict radical and significant innovation. This is because radical innovation assumes a different context and user approach than those of products already on the market. Instead, for radical innovations that aim to create a new meaning, organizations should 'propose' new products and services [22].

An organization's identity can be stated in five categories: the essence of the organization—who you are at your core; the nature of the offerings—what you offer to others; the effects of heritage—where and when you came to be who you are today; the sense of purpose—why you are in business; and the body of values—how your identity is manifested [8]. The true identity of an organization is defined across these five categories. A company can design products or services specifically to create visual brand recognition, so that they carry distinctive references to the 'character' of the brand as manifest in the defined core values. An example of a company that is successful in this is Volvo. Its designs consistently express the core value of safety, along with the typical Scandinavian design that combines functionality and simplicity with beauty and elegance. A typical design element is the strong 'shoulder line' of all Volvo models that makes the sides look more solid and thicker, contributing to the impression of safety [11]. Some organizations are so successful in creating loyalty among their customers that they turn into 'lovemarks'. These brands resonate at an emotional level, and seduce customers beyond reasoning to buy and use their products, often at a premium price [7]. This is an interesting aspect related to transformational innovation, because companies that are able to create such a strong bond with their customers are better positioned to seduce them into changing their behavior. Of course this is only possible if it is done in an authentic way, without violating the trust that consumers put in the brand and the organization.

Sociologists point out that business managers need an understanding of how wider relationships of power and control in society affect, and are affected by, organizations [16]. This relates to the discussion of what is regarded as being part of the organization, and what is not. In this book the organization is regarded as the structure for which the chief executive is directly responsible. It excludes the wider view of the supply chain, customers or other partners with whom contracts are established. These extended views will be discussed in the next section: the ecosystem.

Value for the Ecosystem

The third level of value concerns the ecosystem. Most innovations require more than just one organization and a user group to be successful in the market. Many of today's innovations are not just solitary products, but combinations of products and services within larger systems. In many cases different organizations are involved in these larger systems, which will be called ecosystems in this book. These organizations can be of many different types: profit and non-profit organizations, public and private organizations or industrial or regulatory organizations, to name just a few. Each has a specific role in the ecosystem: for example supplying part of the innovation, advising users, providing maintenance services or approving a new product prior to market introduction. Ecosystems stretch beyond the traditional value chain, the supplier–customer network, or the 'extended enterprise', which also includes suppliers to suppliers and customers of customers. Ecosystems include all stakeholders that have a direct or indirect role in the various phases of the innovation: the definition, creation, realization and extension.

In the industrial economy, the most suitable model was the value chain. But this model no longer fits the new economies of knowledge and transformation. Knowledge, competences and relationships are more important, which is why value networks or value constellations are becoming the dominant model [13]. In such business networks, organizations can achieve greater value than they would be able to achieve on their own. But it requires mutual commitment and mutual dependence to create inter-organizational systems that promote the creation of value [1]. Emergent value networks are formed by negotiation of interests and positions; this is a dynamic process [10]. The actors in this process have to understand not only who is going to contribute what kind of resources and capabilities, but also who will be the provider of specific generic services such as billing, customer care or service management. This is a complex process due to potentially conflicting interests of the organizations involved, which are often from different industries [4]. In the early phases, the process is even more complex due to the inherent uncertainty about user needs and behavior. This uncertainty is particularly high in the case of more radical, transformational innovations, and an iterative process will be applied to elicit the user needs by 'proposing' innovations and improving them where needed. For the actors in the emerging business ecosystem, this uncertainty will limit their ability to secure good positions in the ecosystem right from the start.

The term ‘ecosystem’ stems from biology. It was first used in 1935 to describe a natural unit that consists of living and non-living parts, interacting to form a stable system. The fundamental concepts of ecosystems include the flow of energy via food chains and food webs, and the cycling of nutrients bio-geochemically. Ecosystem principles can be applied at all scales—thus principles that apply to an ephemeral pond apply equally to a lake, an ocean or even the whole planet [17]. The analogy with the business world is recognized by several authors, especially in the knowledge economy. Like business networks, biological ecosystems are characterized by a large number of loosely interconnected participants who depend on each other for their mutual effectiveness and survival. And like business network participants, biological species in ecosystems share their fate with each other. If the ecosystem is healthy, individual species thrive. But if it is unhealthy, individual species suffer deeply. And as with business ecosystems, reversals in overall ecosystem health can happen very quickly. Just like business networks, evolved biological ecosystems—from the Atlantic Ocean to the Amazon—are essentially communities of entities with differing interests bound together in a collective whole [9].

Moore defined a business ecosystem as an economic community supported by a foundation of interacting organizations and individuals—the ‘organisms’ of the business world. This economic community produces goods and services of value to customers, who themselves are members of the same ecosystem. The member organisms also include suppliers, lead producers, competitors and other stakeholders. Over time, they coevolve in their capabilities and roles, and tend to align themselves with the directions set by one or more central companies. Those companies holding leadership roles may change over time, but the community values the function of an ecosystem leader because it enables members to move towards shared visions, to align their investments and to find mutually supportive roles [12].

In the context of transformational innovations the uncertainty is very high, and will be addressed through an iterative process. This also means that the ecosystem will evolve dynamically alongside the creation of the value proposition. Some organizations will exit the ecosystem at certain points because they feel there is insufficient value left for them. Others may enter at a later stage, bringing in new knowledge and experience and thus influencing the value proposition. These dynamics are an inherent part of the process, and should be regarded as necessary to ensure the highest possible value for all members of the ecosystem.

Society is in many cases also a stakeholder that is part of the ecosystem. This can be a specific community such as a neighborhood in which a new, sustainable street lighting system is installed, or it can be society as a whole.

Value for Society

Society is the highest level of value in the model. Users, organizations and ecosystems are all part of society. In many cases the impact of innovations on society is underestimated. Of course, organizations do their best to understand the value of innovations at all levels, and to benefit from it directly or indirectly. But innovations also have their costs, and not all of these costs are taken into account in an integrated model. Indirect costs of producing goods are often not taken into account. Lester Brown gives an intriguing example of what is wrong with the way calculations are made [2]:

> One of the best examples of this massive market failure can be seen in the United States, where the gasoline pump price in mid-2007 was $3 per gallon. But this price reflects only the cost of discovering the oil, pumping it to the surface, refining it into gasoline and delivering the gas to service stations. It overlooks the costs of climate change as well as the costs of tax subsidies to the oil industry (such as the oil depletion allowance), the burgeoning military costs of protecting access to oil in the politically unstable Middle East, and the health care costs for treating respiratory illnesses from breathing polluted air.
>
> Based on a study by the International Centre for Technology Assessment, these costs now total nearly $12 per gallon of gasoline burned in the United States. If these were added to the $3 cost of the gasoline itself, motorists would pay $15 a gallon for gas at the pump. In reality, burning gasoline is very costly, but the market tells us it is cheap, thus grossly distorting the structure of the economy.

In the end, society as a whole suffers from such distortions because in the end the bill is paid by citizens, either by living in a polluted world, or by increased taxes to cover the costs of cleaning up the mess. In this sense citizens are always stakeholders of any innovation, as the ecological footprint of the products and services consumed affects them directly or indirectly. A more integral view on value is needed that will help organizations to create innovations that bring value to users and society. Such an integral view should include potential harmful effects, and support creative processes to reduce harm and increase value.

On the upside, understanding the issues in society also provides a basis for many new transformational innovations. There is plenty of room to improve the quality of life for many people: both in the Western societies, where poverty has mostly been eradicated, although many people live unhealthy lifestyles, and in the developing societies, where the first step is to create a certain level of wealth through innovative solutions that provide income for many.

The next chapter will explore more deeply what value actually is from different perspectives in the social sciences.

References

1. Blankenburg-Holm, Desiree, Kent Eriksson, and Jan Johanson. 1999. Creating value through mutual commitments to business network relationships. *Strategic Management Journal* 20: 467–486.
2. Brown, Lester R. 2008. *Plan B 3.0. Mobilizing to save civilization. Earth policy institute.* London: W.W. Norton & Company Inc.
3. De Mooij, Marieke. 2004. *Consumer behavior and culture: consequences for global marketing and advertising*. London: Sage Publications.
4. De Reuver, Mark, Harry Bouwman, and Timber Haaker. "Capturing Value from Mobile Business Models: Design Issues That Matter." 21st Bled eConference eCollaboration:

Overcoming Boundaries through Multi-Channel Interaction, Bled, Slovenia, June 15–18 2008.
5. Deschamps, Jean-Philippe. 2008. *Innovation leaders: how senior executives stimulate, steer and sustain innovation*. New York: Wiley.
6. Fogg, B.J. "A Behavior Model for Persuasive Design." Persuasive '09, Claremont, California, USA, April 26–29, 2009.
7. Foster, Robert J. 2007. The work of the new economy: consumers, brands and value creation. *Cultural Anthropology* 22(4): 707–731.
8. Gilmore, James H., and B. Joseph Pine II. 2007. *Authenticity: what consumers really want*. Boston, MA: Harvard Business Press.
9. Iansiti, Marco, and Roy Levien. 2004. *The keystone advantage: what the new dynamics of business ecosystems mean for strategy, innovation, and sustainability*. Boston, MA: Harvard Business Press.
10. Ihlstrom Eriksson, C., M. Akesson, M. Bergquist, and J. Ljungberg. Forming a Value Network—Analyzing the Negotiation Between Actors in the e-Newspaper Case. Proceedings of the 42nd Hawai'i International Conference on System Sciences HICSS-42, Big Island, HI, USA, 5–8 January 2009.
11. Karjalainen, Toni-Matti, and Dirk Snelders. 2010. Designing visual recognition for the brand. *Journal of Product Innovation Management* 27: 6–22.
12. Moore, James F. 1996. *The death of competition. Leadership & strategy in the age of business ecosystems*. New York: Harper Business Publishers.
13. Normann, Richard, and Rafael Ramírez. From Value Chain to Value Constellation: Designing Interactive Strategy. Harvard Business Review, July/August 1993.
14. Ohmae, Kenichi. 1988. "Getting Back to Strategy." Harvard Business Review, No. 66, pp.149–156.
15. Oxford Dictionary of Business and Management. ed. Jonathan Law. Oxford University Press 2009. Oxford Reference Online. Accessed 28 December 2010.
16. Oxford Dictionary of Sociology. John Scott and Gordon Marshall. Oxford University Press 2009. Oxford Reference Online. Accessed 8 December 2010.
17. Oxford Dictionary of Zoology. ed. Michael Allaby. Oxford University Press 2009. Oxford Reference Online. Accessed 29 December 2010.
18. Porter, Michael E. 1980. *Competitive strategy: techniques for analyzing industries and competitors*. New York: The Free Press.
19. Reichheld, Fred. 2006. *The ultimate question: driving good profits and true growth*. Boston, MA: Harvard Business Press.
20. Rogers, Everett M. 2003. *Diffusion of innovations*, 5th ed. New York: The Free Press.
21. Treacy, Michael, and Fred Wiersema. 1993. "Customer Intimacy and Other Value Disciplines. Three Paths to Market Leadership" Harvard Business Review, January/February.
22. Verganti, Roberto. 2009. *Design driven innovation*. Boston, MA: Harvard Business Press.
23. Zuboff, Shoshana. "Creating Value in the Age of Distributed Capitalism." McKinsey Quarterly, September 2010.

Chapter 3
Value from Different Perspectives

Etymology

The Oxford Dictionary of English Etymology [59] explains the noun 'value' as: adequate equivalent; material or monetary worth; worthiness. The term stems from the Old French word 'valoir', meaning 'be worth', which is a development from the Latin word 'valere', meaning 'be strong, be worth'. From the fourteenth century onwards it is also used to indicate a relative status or estimate, and from the sixteenth century onwards as an amount represented by a symbol. The derivations 'valuable' and 'valuation' came into use in the sixteenth century.

The Oxford Dictionary of English [58] defines 'value' as: the regard that something is held to deserve; the importance, worth or usefulness of something (for example: your support is of great value); the material or monetary worth of something (as in: prints seldom rise in value; or, equipment is included up to a total value of $500); the worth of something compared to the price paid or asked for it (the wine represents a good value for $17). 'Value for money' is used in reference to something that is well worth the money spent on it (such as: this camera is really good value for money). The verb 'to value' is defined as: estimate the monetary worth of; consider someone or something to be important or beneficial; or have a high opinion of (e.g. she had come to value her privacy). The term value is also used in algebra, music and linguistics.

The term 'values' is defined as: principles or standards of behavior; one's judgment of what is important in life (for example: they internalize their parents' rules and values).

What is interesting is the difference between 'value' and 'values'. 'Value' refers to the value of something, whereas 'values' refers to what is important in life. The latter is therefore a more comprehensive, holistic view. In the context of innovations that aim to provide value through transformations, both notions are important. The innovation itself should be of value not only to stakeholders, but it should also contribute to their values. A computer game to stimulate the physical activity of children, for example, should be of value to the children in the sense

E. den Ouden, *Innovation Design*, DOI: 10.1007/978-1-4471-2268-5_3,

that they love to use it. At the same time the game addresses the values of improving quality of life, by evoking a healthier lifestyle through the stimulation of the activity of the children.

The Philosophical View

Philosophers have focused their discussions on value on three issues, which are connected to each other [64]:

- on what sort of property or characteristic of something its 'having value' or 'being of value' is based;
- on whether having value is an objective or subjective matter, whether value resides in the object or is a matter of how we feel toward it;
- on trying to say what things have value, are valuable.

The value of a product or service is not a property that can be directly measured scientifically, like other properties such as weight or volume. It seems to be more a relational property: the value only exists in relation to a human. Something could thus be called 'valuable' insofar as we cherish it, although common sense might well say that this is the wrong way round: we cherish it because it is valuable [64]. Consider, for example, the value of a piece of jewelry. A golden ring set with a beautiful diamond has objective value because it is made of precious materials. The objective value of a silver ring set with a pearl is less because it is made of less valuable materials. However, the silver ring could be of much greater value to a specific person because it belonged to her grandmother and has been in the family for a long time. The subjective value of the silver ring can therefore be higher than that of a golden ring that has no history for the person concerned.

The differences in the objective view of 'having value' and the subjective view of 'being of value' can also be found in the way economists or sociologists approach value, as will be explained in the following sections.

In the context of innovation it is important to consider both the objective and subjective value of products or services, as these may influence the decisions people take and how they behave. People may cling to an object and keep on using it, even if a better alternative is available. For example, the use of a percolator to make coffee has proven to be unhealthy because of the resulting higher cholesterol levels. Other methods of brewing coffee are healthier. But if the percolator has emotional value a person may not want to part with it, even if he or she agrees that other coffee tastes better and is healthier. Similar things happen with old cars. Although inherently less safe, and certainly not beneficial for the environment, many people love to drive classic cars. Understanding what things are valued by people, and why, is essential for understanding their feelings toward new solutions, and what actions can be expected.

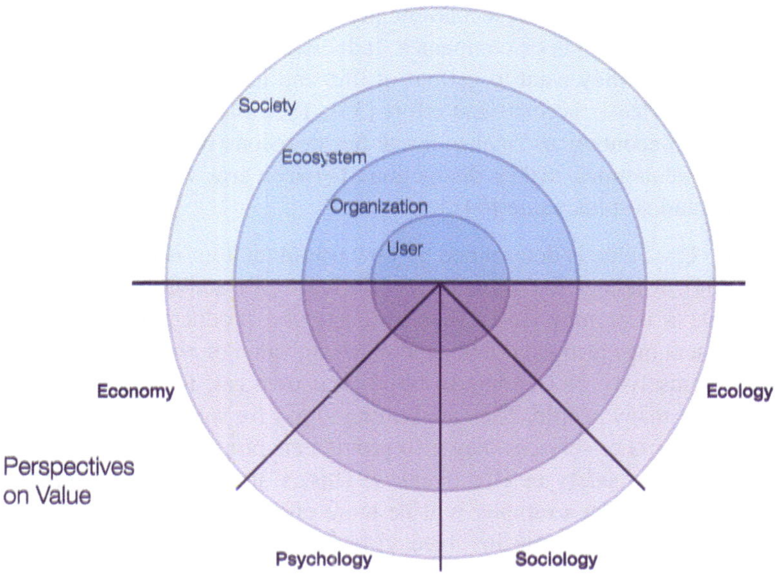

Fig. 3.1 Integrating views on value from social sciences in the value framework

Integrated View

In the subsequent sections the term 'value' will be explored from the perspective of the social sciences that provide a relevant context for innovation. The different views on value will be integrated into one overall Value Framework. This combines the value concepts from economics, psychology, sociology and ecology, and explains them at the four levels of value described in the previous chapter: user, organization, ecosystem and society. Figure 3.1 shows the structure of the Value Framework that will be used to integrate the different perspectives into one comprehensive view on value: each of the social sciences—economics, psychology, sociology and ecology—is indicated on a radius.

The Economic View

Economics is the branch of knowledge concerned with the production, consumption and transfer of wealth [53]. It studies human behavior in relation to scarce means that also have alternative uses [74]. Economics distinguishes microeconomics, in which the unit of analysis is an individual such as a household or a company; and macroeconomics, in which the unit of analysis is the economy as a whole. Although originally the definition of economy by Aristotle was defined as: 'the art of living and living well', the general focus in the past decades has been reduced to 'the art of money making' [14].

Microeconomics In the microeconomic perspective the focus is at the individual level, such as households or companies. Individuals are assumed to have a fairly clear idea of what they want to get out of life, and to be trying to get as much as possible for the least sacrifice and effort [33]. This assumption leads to the idea that the 'homo economicus' makes his or her decisions rational.

Classical economics makes distinctions between use value, exchange value, labor value and surplus value [64].

Use Value Use value is determined by how useful an item is to a given person or situation [64]. It relates the value of an offering to the activities of the user [71]. For example a user may save labor by using the product, as with a washing machine. Users may also experience the relieving value of not having to do certain activities themselves. Not having to build your own car, for example, may be a great relief for many people. They are willing to pay for someone else to do the job for them. Products or services may also provide enabling value through enhanced ease, productivity, safety or effectiveness. This is often the case in professional services. A lawyer, for example, will be more effective in defending a case than someone else who is getting involved in legal proceedings for the first time. A related concept to use value is utility value, which addresses efficiency, durability, time management, accessibility, appropriateness, compatibility, safety and economy of using products or services [5]. In the business-to-business sector, for example, customer productivity is a criterion that becomes more important as a means to provide added value [71]. Companies selling medical equipment for use in hospitals should make sure that the efficiency of the medical staff is improved as well. When providing a product that helps them to do the job well, such as an X-ray scanner producing high-quality images, they should understand the workflow in the hospital and contribute to improving efficiency, such as by enabling electronic data management of the scanner images.

The use value of products has always been a central theme in design. A designer directly influences the use value of a product [5]. The physical properties of the product, such as its size and shape, can make or break its usability. Most people have experience with teapots or water cans that look great, but drip when used. Poor design can reduce the use value of a product, while good design can greatly enhance it. Especially in more complex products with numerous functions and features, good design can make a big difference. Good design can go as far as to create a pleasurable experience, through which people sincerely enjoy using the product. Apple is well known as a company that has done just with products such as the iPhone and iPad.

In line with the earlier philosophical discussion on value, a similar discussion can be held on the question of whether products fulfill needs. Keinonen emphasizes that artifacts seldom fulfill needs, but enable people to take the necessary actions to satisfy their needs [41]. It is therefore important to understand the needs of users. Understanding users' fundamental needs requires taking up a holistic non-reducing view of users and allowing them to behave autonomously in the roles and situations they have assumed [41]. Traditional ways of looking at users have all too often ignored this. In many cases, the use of the product was studied out of the normal context of use; for example in a usability lab. This leads to unrealistic conclusions on the use value of products. In the case of televisions, we might test how easy people find the initial setting of the channels. The test might include people who would never do such a task themselves, perhaps because they would ask someone else to do it. Asking subjects like these might make them feel clumsy and ignorant, and the results of the test would not be representative of reality. For these people, other tasks like showing a short movie clip of the children to friends could be much more important. And for them to appreciate a new television model, all they need to do is complete these tasks to their satisfaction. It is important to understand the autonomous behavior of people in the different roles they have in their lives—as a family member, as friend, as a professional or as a club member. Value is the result of interaction of a user with a product or service, in which a product's value pertains to the experience as associated with that product [5]. Thus although managers and designers can influence the value of products and services, value is subjective and determined by users, and it would be wrong to think that value is determined by management or design decisions [36].

Defining use value in this way, including the subjective and intangible concept of 'experiences', goes beyond its traditional and more functional definition in economics.

Exchange Value Exchange value reflects the price of an item on the market [63]. It refers to the idea that the value of a product or service can be compared to the value of another product or service. Money is the usual asset offered in exchange, either as a physical object (cash) or more abstract (credit). Money is unique because it can indicate exactly how much more one thing is worth than another [33]. The value of a product or service is then defined as the greatest amount of money that consumers are prepared to pay for it [51]. In a wider definition proposed by Pitelis, value is the perceived worthiness of a subject matter to a socio-economic agent who is exposed to and/or can make use of the subject matter in question [67]. Worthiness encompasses rarity, aesthetic appeal, satisfactory price or value for money.

Within the economic view, marketing scholars have been at the forefront of attacking the view that transactions are made based on rational calculation [35, 53]. People may miscalculate the expected costs and benefits due to the lack of perfect information or information overload, and therefore making an imperfect analysis [53]. Especially in more complex situations the 'foreseeability' may be limited. For a simple product with a relatively short period of use and after many repeated purchases, it might be possible for a customer to rationally assess the value and the price, and to judge whether the product is providing the expected value for money. But this is much harder for more complex products or those with a longer period of use. Rational decisions are also much harder when more intangible aspects like aesthetics are involved.

Exchange value is important for innovations, as it determines the financial feasibility of the innovation in the long run. For transformational innovations, it is important to see which of the stakeholders would be willing to pay for what part of the solution. For a home healthcare service it might be not just the user or the patient, who is willing to pay, but the other stakeholders as well. The family members of the patient might be willing to pay for a service that provides them with the comfort of being assured that their family member is being taken care of well. An insurance company might be interested because it would reduce the total costs, as home healthcare could be more cost-effective than staying in a hospital. The business model should therefore be built on the interests of the various stakeholders and the related exchange values.

Labor Value Labor value reflects the amount of human effort invested in the production of a good [63]. Karl Marx emphasized that the value of a commodity is derived from the human labor that went into producing it, but that this is generally forgotten when the object is bought. It seems that the value is derived directly from the qualities of the object itself [48].

Different organizations contribute to the value of products in value chains. As Porter indicated, organizations create value for consumers across a range of primary and supporting activities [69], including production, distribution, marketing and sales.

In this view it looks as though customers destroy value. They are right at the end of the value chain, and do not add any further value to the product itself. This destruction of value becomes apparent through the depreciation of the value of goods during use that is common in business economics thinking.

In the context of innovation, the concept of labor value has a direct link to the costs of developing and producing a product or service, but it relates less directly to the value of an innovation to customers.

Surplus Value Surplus value reflects the difference between the price of a good over the costs of labor and raw materials [63]. Essentially surplus value exists if the price of the product or service can be set at a higher level than its costs, in other words when a margin can be generated. It requires an understanding of the value in the eyes of the customer. This applies not only to individual goods and services, but also to those that are integrated in solutions. For a customer, the value of a solution is the sum of the value of the individual products and services, the value of marketing and operational integration and the value of customization for the customer's specific needs and context [77]. Surplus value implicitly addresses the difference between the objective, measurable value of a good (costs) and the price a customer is willing to pay based on subjective, not directly measurable value.

Surplus value has the tendency to decline over time. While an innovative product or service can usually be sold at a higher price that price—and with it the margin—falls over time as competition increases. This effect drives companies to seek new ways to differentiate their offerings from those of their competition, aiming to avoid the commoditization trap. Pine & Gilmore explain this process through the example of a true commodity: the coffee bean [66]:

> Companies that harvest coffee or trade it on the futures market receive a little more than $1 per pound, which translates into one or two cents a cup. When a manufacturer grinds, packages, and sells those same beans in a grocery store, turning them into a good, the price to a consumer jumps to between 5 and 25 cents a cup (depending on the brand and package

size). Brew the ground beans in a run-of-the-mill diner, corner coffee shop, or bodega and that service now sells for 50 cents to a dollar per cup. So depending on what a business does with it, coffee can be any of three economic offerings—commodity, good, or service—with three distinct ranges of value customers attach to the offering. But wait: Serve that same coffee in a five-star restaurant or espresso bar, where the ordering, creating, and consumption of the cup embodies a heightened ambience or sense of theatre, and customers gladly pay anywhere between $2 and $5 for each cup. Businesses that ascend to this level of value establish a distinctive experience that envelops the purchase of coffee, increasing its value (and therefore its price) by orders of magnitude over the original commodity. Or more. Immediately upon arriving in Venice, Italy, a friend asked a hotel concierge where he and his wife could go to enjoy the city's best. Without hesitation they were directed to the Café Florian in St. Mark's Square. The two of them were soon at the café in the crisp morning air, sipping cups of steaming coffee, fully immersed in the sights and sounds of the most remarkable of Old World cities. More than an hour later, our friend received the bill and discovered the experience had cost more than $15 a cup. Was the coffee worth it, we asked? "Absolutamente!" he replied.

A progression of economic value can be seen in the subsequent offering of commodities (extracted from the earth, exchanged as raw, fungible offerings), to goods (tangible things manufactured from commodities), to services (intangible activities delivered on behalf of an industrial customer), to experiences (memorable events engaging individuals in a personalized way) and to transformations (outcomes that guide customers to change themselves) [31].

Macroeconomics Macroeconomics is the branch of economics that analyzes the national economy at an aggregate level, and is concerned with national income: the income accruing to a nation by virtue of its productive activities. In measuring national income, one of the common approaches is the Gross Domestic Product (GDP)—the unduplicated value of all goods and services produced in the domestic economy [64]. Although addressing macroeconomic issues is outside the influence of a single innovation, more socially directed innovations do attempt to address solutions that improve the wealth of larger groups of people [96]. The so-called 'bottom-of-the-pyramid' projects attempt to alleviate the poverty of the billions of people living on less than $2 per day [70].

Wealth In the classic 'Wealth of Nations', Adam Smith argued: "Every individual necessarily labors to render the annual revenue of society as great as he can. He generally neither intends to promote the public interest, nor knows how much he is promoting it. He intends only his own gain, and he is, in this, as in many other cases, led by an invisible hand to promote an end which was no part of his intention." [83]. His theory of the 'invisible hand' emphasizes that if everybody pursues his own interests, all activities will eventually benefit society.Wealth can be considered at both individual and societal levels. An individual who has an abundance of items of economic value, such as money, real estate or goods, is considered wealthy. A wealthy society is one that has more resources than a poor one. The concept of wealth is relative and varies between societies, and between groups of people within society, and it changes over time. A person owning $5,000 in cash might be considered rich in an impoverished nation, but would not be considered as very wealthy in one of the Western European countries.

In line with Smith's concept of the 'invisible hand', Gilmore and Pine argue that wealth is maximized when individuals specialize in what they do best and then trade with other individuals for what they do best. Money is the means by which this is made easier: one can choose to trade the fruits of his own labor with those of another person [31]. In the current, networked economy, there are new opportunities for people to contribute from their own strengths. As Friedman describes it: 'the world has become flat'[29]. Technology and the globalization in the market have made it possible for India and China to become part of the global supply chain of services. Many jobs in the knowledge economy can be done regardless of place as long as one is connected to the Internet. This has already led to an explosion of wealth in the middle classes of the world's two largest nations.

Gross Domestic Product The Gross Domestic Product (GDP) is defined as the total amount of goods and services produced by a country annually. It does not include income from investments or overseas possessions. GDP gives an indication of the strength of national industry [64]. It was already argued in the 1970s that GDP, with its focus on production, overlooked the fact that irreplaceable capital is consumed, namely that which man has not made, but is simply found to be available on earth [78]. Every economist and business person is supposed to be familiar with the distinction between income and capital; however, this concept is generally not applied at the macroeconomic level. As a result, as well as the 'externalities' that will be discussed in the next paragraph, such as pollution, there is the problem of the depletion of 'natural capital' and natural resources.

The concept of Gross National Happiness (GNH) was developed in an attempt to define an indicator that measures quality of life or social progress in more holistic and psychological terms than GDP. The term was coined in 1972 by Bhutan's former King Jigme Singye Wangchuck [54], and has been further developed and gained attention through international conferences. This concept is much broader than the economic perspective, and will be discussed in more detail later.

Externalities There can be harm or costs to parties or individuals who are not involved directly in the exchange, or costs and/or benefits that fall outside the price system [53]. An example of these 'externalities' that affect individuals as well as society is pollution (a non-monetary effect). These externalities can be the unintended consequences of miscalculations or unawareness, but they may also result from more or less conscious decisions to disregard them due to self-interest.

For innovations it is important to understand the limitations of the economic focus, to make sure that potential externalities are internalized in the market mechanism, and to take into account the wider definition of worthiness. By choosing the boundaries of the considered ecosystems, we can ensure that the 'externalities' become part of the system.

Table 3.1 Value definitions and concepts from an economic perspective

	User	Organization	Ecosystem	Society
Value definitions	Use-value Exchange value	Exchange value Labor value Surplus value	Exchange value Surplus value	Wealth GDP
Value concept	Value for money	Profit	Stability	Prosperity

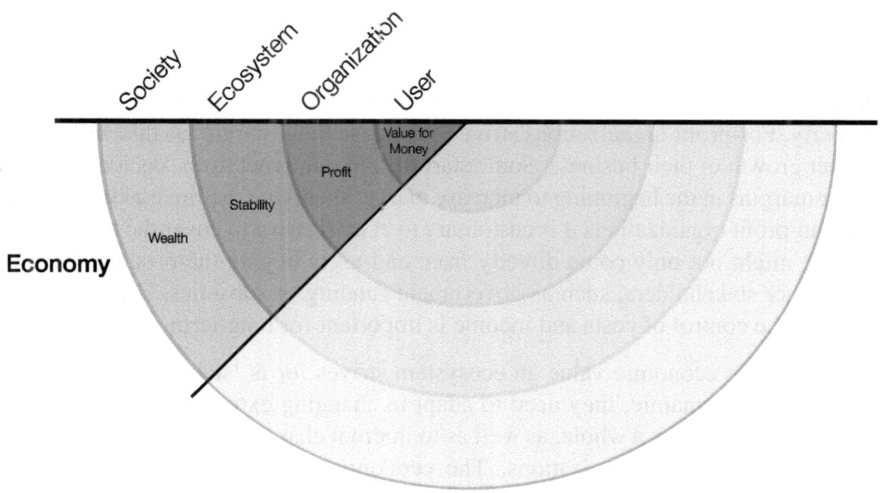

Fig. 3.2 Economic value concepts in the value framework

Economic Value Concepts

The value definitions discussed above apply to one or more of the levels of value presented in the previous chapter. Obviously the microeconomic definitions fit more with the levels of customer and organization, while the macroeconomic definitions fit with society as a whole. Table 3.1 gives an overview of the different definitions and their applicability at the four levels.

To make things practical, an overall concept is identified for each level that can be applied to address the values at that level. The economic value concepts are part of the Value Framework, as shown in Fig. 3.2. A more detailed description of what is meant by these concepts is given in the following sections.

Value for Money The economic value of a product or service for the user can be defined as 'value for money'. The user, or customer, will strive to achieve the maximum exchange and use value for his money. In the eyes of a user or customer, this also includes to intangible value such as that of an experience. People act in ways that seem economically irrational, this is only because the values they are maximizing are not material [33]. They may take decisions that are not based on a calculating, purely rational reasoning, for example by deciding intuitively what provides value for them in a particular context. This will be discussed in more detail in the sections on psychology and sociology.

Profit The economic value that companies strive for can be summarized as 'profit'. Profit is essentially the difference between the cost incurred in developing, producing and delivering the product or service (man, machine, material), and the income the organization can gain from the product or service. In most business situations this income is the price for which the product or service can be sold on the market: the exchange value, or the amount of money that customers are willing to pay for the product or service. Organizations can have other income as well, either directly or indirectly related to the innovation, such as income from licensing intellectual property. For-profit organizations strive to increase their margin, as this will enable further growth of their business. Some start-ups, like Internet firms, decide to accept lower margins in the beginning to increase market share first, before making a profit. For non-profit organizations it is customary to at least strive to cover the costs. Their income might not only come directly from end-users buying the product, but also from other stakeholders, such as government funding or subsidies. For each organization the control of costs and income is important for long-term survival.

Stability The economic value an ecosystem strives for is 'stability'. Ecosystems are inherently dynamic: they need to adapt to changing external conditions in the market and society as a whole, as well as to internal changes such as the exit of one of the participating organizations. The economic value of an ecosystem is to provide financial stability for its stakeholders by adapting to new conditions, and thereby providing sustainable value delivery for the stakeholders. The resilience of an ecosystem is a key success factor in the long run.

Wealth Economic value for society can be summarized as 'wealth'. A flourishing society is a rich society built on economic activities that result in a high Gross Domestic Product. Wealth is maximized when individuals specialize in what they do best and then trade with other individuals for what they do best [31]. A wealthy society is able to provide its citizens with good healthcare, education and infrastructure, thereby contributing not just to the welfare but also to the well-being of the people. Economic value is an enabler for psychological, sociological and ecological value.

The Psychological View

Psychology is concerned with the study of the human mind and its functions, especially those affecting behavior in a given context [58]. Value in the perspective of psychology is a desirable trans-situational goal, serving as a guiding principle in people's lives [43, 75, 79]. Traditionally psychology has mostly been concerned with negative aspects like depression. Positive psychology is a relatively new science that is a counter-reaction to this focus on the negative side, and specifically considers what makes life worth living [81]. Terms used in this respect are subjective well-being, happiness and life satisfaction. Values in the psychological perspective define what people strive for—the human values—as well as how these influence their behavior—the motivational values.

Subjective Well-being and Happiness Subjective well-being concerns the assessment of people's evaluative reaction to their lives and societies: their normative ideals, subjective experiences and ability to select the goods and services they desire [22]. How people assess their lives depends on wealth and culture. As the assessment is subjective, people relate their evaluation on what is common around them. Cultural norms dictate how they feel in certain circumstances [21].

It may look as if improving subjective well-being is beyond the influence of innovation, but well-being set points can be changed [20]. Innovations can aim to provide pleasant experiences and make people more aware of them. People who have experienced a series of bad events rate their happiness lower, but after a positive event they increase their scores. Individuals who regularly reappraise their happiness tend to experience more positive emotions and fewer negative ones. Among older people the endorsement of coping styles such as using humor, seeking information and 'keeping going' predicts adjustment at old age.

People judge their global life satisfaction differently for different life domains such as family life and working life. When people judge their life satisfaction, they do this both through an objective evaluation and a subjective appraisal [8, 91]. The objective evaluation is cognitively guided and assesses the degree to which life fits their current values and ideas. The subjective, intuitive affective appraisal assesses the degree to which the individual enjoys life. Contentment is the result of the comparison of 'life as it is' with standards of 'how life should be' [92].

Veenhoven argues that the best way to measure values is to measure happiness, as this is a more inclusive merit than most other values [91]. It reflects an optimal combination of other values, taking into account that there are trade-offs between values. He also emphasizes that people judge tangible matters through a comparative evaluation, and intangible matters by how they feel about them. This is an important notion because it means that both forms of judging should be taken into account when striving to improve happiness through innovation.

The ultimate value is the greatest happiness of the greatest number of people, best measured by how long and happily people live [93].

Human Values Human values are defined as core conceptions of what is desirable within every individual and society. They serve as standards or criteria to guide not only actions but also judgment, choices, attitudes, evaluations, arguments, exhortations, rationalizations and attributions of causality. The individual's value orientation influences the judgment of global life satisfaction [55]. The number of central human values is small, and they are comparable all over the world. They are the results of societal demands and psychological needs, and are learned and determined by culture, society and personal experience [75].

Human values are standards that are to a large extent derived, learned and internalized from society and its institutions. These standards guide the development of a socially defined sense of self as a competent and moral member of

Table 3.2 Human values as defined by Rokeach [75]

A comfortable life—a prosperous life	Inner harmony—freedom from inner Conflict
An exciting life—a stimulating, active life	Mature love—sexual and spiritual intimacy
A sense of accomplishment—lasting contribution	National security—protection from attack
A world of peace—free of war and conflict	Pleasure—enjoyable, leisurely life
A world of beauty—beauty of nature and the arts	Salvation—saved, eternal life
Equality—brotherhood, equal opportunity for all	Self-respect—self-esteem
Family security—taking care of loved ones	Social recognition—respect—admiration
Freedom—independence, free choice	True friendship—close companionship
Happiness—contentedness	Wisdom—mature understanding of life

society. Mass media enable people to become aware of themselves with respect to their stance on issues such as racial justice, sex equality, involvement in an unjust war or threats to health from pollution, and to make possible changes in the way they perceive themselves.

The most widely used model of human values is the one defined by Rokeach in 1979; his list of values is shown in Table 3.2.

Not all of these values can be addressed by innovative products or services, but they can certainly be an inspiration for meaningful innovations. Moreover, they can serve as a checklist to evaluate innovative ideas for their potential positive and negative impact on one or more values, either directly or indirectly.

Motivational Values A major goal of research on values has been to relate individual differences in value priorities to differences in attitudes and behavior. Schwartz has carried out extensive research to derive a nearly comprehensive set of different motivational types of values that are recognized across cultures. Values can be distinguished by the type of motivational goal they express. People have three universal requirements of human existence: biological needs, requisites of coordinated social interaction and demands of group survival and functioning. Groups and individuals represent these requirements cognitively as specific values about which they communicate to explain, coordinate and rationalize behavior [79].

Table 3.3 shows the ten motivational values that Schwartz derived from his research. The table indicates the central goal of each of the values, and in brackets the specific values that primarily represent the central goal.

The specific values that primarily represent the central goal are shown in brackets.

These ten motivational values are also related to each other. Actions taken in pursuit of each type of values have psychological, practical and social consequences that may conflict or may be compatible with the pursuit of other value types. The total pattern of relations of conflict and compatibility among value priorities yields the structure of value systems represented in Fig. 3.3. Competing values emanate in opposing directions from the center; complementary types are in close proximity going around the circle.

Understanding the motivational values is important for transformational innovations. People are not likely to change their behavior in a direction that is not in line with their value orientation. As the motivational values may conflict, it is even

Table 3.3 Motivational types of values as defined by Schwartz [79]

Central goal	Specific values
Power	Social status and prestige, control or dominance over people and resources (social power, authority, wealth, preserving public image)
Achievement	Personal success through demonstrating competence according to social standards (successful, capable, ambitious, influential)
Hedonism	Pleasure and sensuous gratification of oneself (pleasure, enjoying life, self-indulgent)
Stimulation	Excitement, novelty, and challenge in life (daring, a varied life, an exciting life)
Self-direction	Independent thought and action-choosing, creating, exploring (creativity, freedom, independent, curious, choosing own goals)
Universalism	Understanding, appreciation, tolerance and protection for the welfare of all people and for nature (broad-minded, wisdom, social justice, equality, a world at peace, a world of beauty, unity with nature, protecting the environment)
Benevolence	Preservation and enhancement of the welfare of people with whom one is in frequent personal contact (helpful, honest, forgiving, loyal, responsible)
Tradition	Respect, commitment and acceptance of the customs and ideas that traditional culture or religion provide the self (humble, accepting my portion in life, devout, respect for tradition, moderate)
Conformity	Restraint of actions, inclinations, and impulses likely to upset or harm others and violate social expectations or norms (politeness, obedient, self-discipline, honoring parents and elders)
Security	Safety, harmony and stability of society, of relationships, and of self (family security, national security, social order, clean, reciprocation of favors)

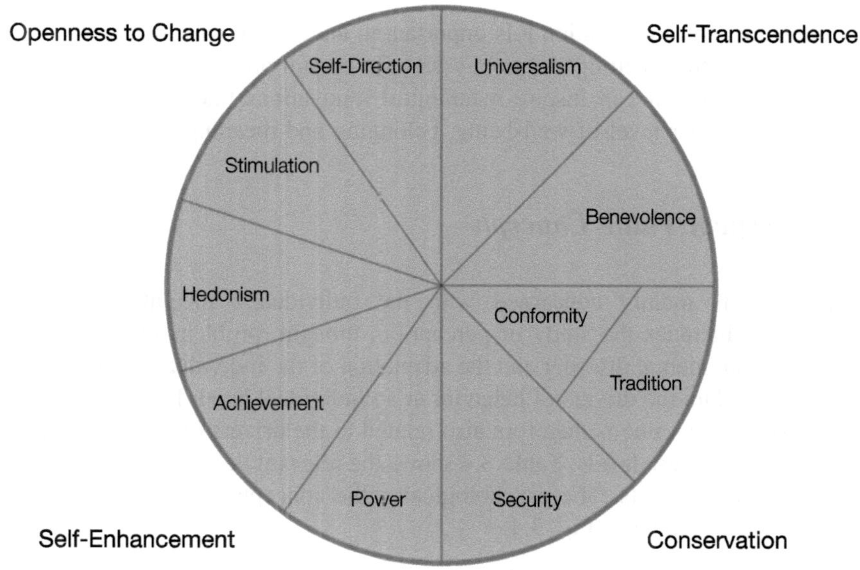

Fig. 3.3 Theoretical model of relations among the ten motivational types of values [76, 80]

more important to understand how people with different values perceive a specific innovative solution. One person might discard what may be a good fit for another. People who value universalism are more willing to adopt innovations that are beneficial to the planet than people who look for social status. The motivations of the target users should be understood well for an innovation to become successful.

Positive Psychology Positive psychology is the psychology of valued psychological phenomena, such as contentment and satisfaction (with the past), hope and optimism (for the future) and happiness and well-being (in the present). It focuses on positive subjective experiences, such as love and aesthetic appreciation; positive individual qualities, such as courage, interpersonal skill, perseverance, originality, and wisdom; and positive social phenomena, such as cooperation, altruism and tolerance [64]. It started as a reaction to the almost exclusive focus on pathologies that arise when life is barren and meaningless, which had been common in psychology for many decades.

Seligman [82], who introduced the term positive psychology, defines 'a life worth living' at three levels. These levels are similar to the classic distinction between 'having', 'loving' and 'being' [1].

A pleasant life—a life of enjoyment: experience, forecast and savor positive feelings and emotions that are part of normal and healthy living, e.g. relationships, hobbies, interests and entertainment;
A good life—a life of engagement: beneficial effects of immersion, absorption and flow that individuals feel when optimally engaged with their primary activities, experiencing a positive match between strengths and task;
A meaningful life—a life of affiliation: positive sense of well-being, belonging, meaning and purpose from being part of and contributing to something larger or more permanent than themselves, e.g. nature, social groups or a movement.

In the context of innovation it is important to understand what drives people, as this influences their willingness to buy and use certain products and services. But more importantly, it can inspire meaningful innovations that support people in achieving a higher level of well-being, belonging and meaning.

Psychological Value Concepts

Psychology is mainly concerned with the individual's mental activity and behavior. It includes the study of perception, thought, problem-solving, personality, emotion, mental disorder and the adaptation of the individual to society [64]. It aims to explain the drivers of behavior at an individual level. Its applicability to the four levels of value is therefore also related to the drivers of the people taking the decisions at these levels. Table 3.4 shows the relevant definitions as well as the derived value concepts. The psychological value concepts are part of the Value Framework, as shown in Fig. 3.4.

Happiness As indicated above, happiness is the most inclusive merit to measure an individual's psychological well-being. 'Happiness' has therefore been selected

Table 3.4 Value definitions and concepts from a psychological perspective

	User	Organization	Ecosystem	Society
Value definitions	Human values	Organizational values	Organizational values	Subjective well-being
	Motivational values			Happiness
	Happiness			Contentment
	Contentment			
Value concept	Happiness	Core values	Shared drivers	Well-being

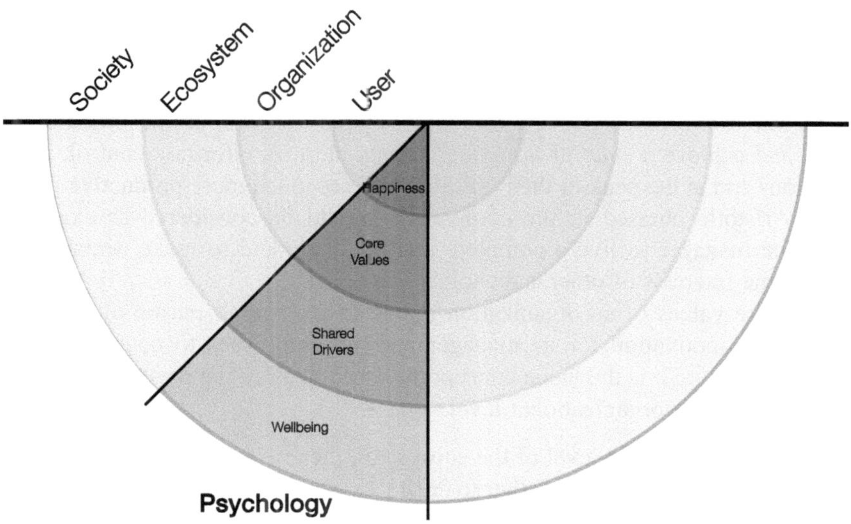

Fig. 3.4 Psychological value concepts in the value framework

as the concept to represent psychological value at the level of the user. At a deeper level, human values and motivational values explain the behavior of people. Insights into the personal drivers of users' lives can help to improve innovative ideas, or inspire new ideas to improve the happiness of specific users in specific contexts.

Many innovations will only offer pleasure, partial satisfaction (related to a part of life) or peak satisfaction (temporary satisfaction). But the ultimate aim of transformational innovations is to improve the overall life satisfaction of users. The notion that satisfaction measurements include both an objective, cognitive evaluation and a subjective, intuitive appraisal is useful in the evaluation of a user's happiness with an innovative solution. Both aspects need to be checked when proposing innovations to prospective users.

Uses will eventually change their behavior when new opportunities for reaching prioritized latent values are available [86]. This aspect is important for transformational innovations: by addressing the values of users, and being able to communicate this, prospective users can be expected to adopt the innovation and use it extensively, thereby increasing their overall happiness or life satisfaction.

Core Values The purpose and objectives of an organization are encapsulated in a mission statement, which is used to communicate this purpose to all stakeholder groups, both internal and external, and to guide employees in their contribution toward achieving it. Organizational values are designed to guide the behavior and thinking of employees in everything they do [62]. Values are typically embodied in slogans and symbols within the organization.

Many companies understand the importance of employee engagement for achieving the organization's ambitions. Innovations can directly contribute not only to the drivers of users (as stated in the previous section), but they can also indirectly contribute to the drivers of employees. If employees are involved in the definition, creation or realization of meaningful innovative solutions, their motivational values are addressed. Providing employees with opportunities to add value by contributing to a meaningful task has also proven to be a way to increase job satisfaction and reduce the risk of burnout [90]. Moreover, employees who are absorbed in their work, and achieve a state of flow, feel strong, alert, in effortless control, unself-conscious and at the peak of their abilities, and are much more productive [15].

As Aristotle phrased it: the corporation should be considered as existing to allow the manager to live a complete and good life and to make decisions that involve the interests of other stakeholders [95].

The core values of an organization represent the overall reason of being, and provide the motivation for its management and employees to contribute to the creation of value. It is the value concept that encompasses the psychological value perspective at the organizational level.

Shared Drivers At the level of the ecosystem, there is no single decision-maker. The ecosystem itself is put together to create a system to deliver new benefits to its members [52]. To be successful in this pursuit, it is important that the members of the ecosystem share the mission of the ecosystem as a whole. All the members will have their own core values, as defined in the previous section, but together they have to share an overall mission for the ecosystem. It also means that they have to agree on the main drivers that rule the decisions and direct the behavior of the various members of the ecosystem. Most often, this sharing and alignment of drivers is not done explicitly, but happens over time as people collaborate.

Ecosystems adapt dynamically to external and internal events, and over time the mission of the ecosystem may also evolve. This may mean that some of the members no longer feel comfortable in the system, and may leave it. New members can also enter it at a later stage, contributing to new insights and motivation for the system as a whole. It is important to understand these dynamics, and the role of the ecosystem in providing stability and a sense of direction for its members, while at the same time ensuring the prolonged life of the ecosystem itself by adapting to the ever-changing condition. The shared drivers are therefore not a fixed set, but will to some extent be adaptive.

An example of an ecosystem with shared drivers is 'Zorg Binnen Bereik' (translated as 'care within reach') in the Netherlands, which was started as an independent foundation by the industrial company Philips Healthcare and the health insurance company Achmea. The aim of the foundation is to invest in the development of innovative care solutions that aim to improve the independence of patients with chronic diseases, thereby improving their quality of life and reducing the number of hospital visits required. The ecosystem of the foundation includes solution providers such as Philips, insurance companies such as Achmea, care providers, healthcare professionals, medical staff and patient associations. What binds them is the shared motivation to improve the independence of the patients. This even makes it possible for competitors to collaborate. Each member of the ecosystem maintains its own core values, but the shared drivers make it possible to collaborate for a greater cause.

Well-Being The overall value for society is well-being. The term well-being came into use in contrast to the idea of simple 'economic welfare' [93]. Psychological well-being of people is very much influenced by society, because society provides the reference for people to judge their individual situation. Through accepted values and beliefs about what is considered important, cultural aspects have a significant influence on how people perceive their own well-being. This aspect will be discussed in more detail in the following section on the sociological perspective. The prosperity of a society also influences the perception of well-being. Hence there is a strong link between the societal level and the individual level of the user. Society provides the bigger picture and the reference framework for individuals to judge their own personal happiness.

Although the idea of improving the well-being of larger groups of people through new products and services may at first glance seem far-fetched, there are quite a number of interesting examples that have an impact on societal well-being. One of these is the microcredits scheme provided by the Grameen Bank, which has improved the well-being of 80% of the families in Bangladesh. This improvement is partially economic, by enabling them to have a better income. But it has also provided women who have benefited from microcredit to build their businesses with a sense of accomplishment, family security, freedom, self-respect and social recognition. Being able to address these human values on such a scale increases the well-being of a large community: Bangladeshi society as a whole.

The Sociological View

Sociologists and anthropologists often have a completely different understanding of value from that of economists and psychologists. They generally speak of social or cultural values as mechanisms of solidarity and collective identity. The focus is on groups of people and relationships. The sociologist Talcott Parsons based his influential account of the dynamic between solidarity and heterogeneity around the notion of a deep level of shared values that permits relatively wide variance at the level of attitudes and ideas. The question of socialization, or the reproduction of values—fundamental to any account of the persistence of a society or culture over time—has also been central to modern sociology and anthropology [63].

In a social context, the roles of individuals are important. We are what we are perceived to be by others, but they might only see us in a certain role. In general people tend to see only a partial perspective; there appears to be a human limit to understand that one's own perspective is only part of a much larger totality. This hinders the human imagination in social matters [33]. A person may have many roles, e.g. as a lawyer, as an employee of a law firm, as a husband, as a father of his children, as a fanatical tennis player, as the member of a political party, as the volunteer secretary of an interest group, as a neighbor, as a voting citizen etc. People may know the person in only one role, tending to see that as the whole person and unknowingly overlooking the other roles. However, the perception of what this person values may be different in the different roles and the related communities.

Sociology and anthropology have contributed with many insights into value and how it is perceived by different groups of people. The important views are summarized in the following sections.

Absolute, Transcendental Value Objects can have transcendental value. Examples of objects that have a high transcendental value are the crown jewels of the United Kingdom. Nobody would ever give these away. Still they could be lost, stolen, forgotten or destroyed. Their value can best be measured in terms of the fear of loss [33]. As they will never be put up for sale, their value cannot be measured by how much anyone is willing to pay for them. Instead, their value is measured by how little one would wish to give them up. Objects of transcendental value are the very last things anyone would want to part with.

The absolute value of an object can be defined as an effect of all the efforts people have made to maintain, protect and preserve it [33]. In a way this is similar to the definition of labor value, but it has a broader scope: it also includes the activities to protect and preserve the object (which is not seen as labor in the Marxist view). The only problem with this measure is that it is very hard to count: nobody keeps track of those efforts, in contrast to labor that is recorded meticulously by companies that make products.

Comparing the values of different objects requires some initial grounds of similarity. If objects cannot be compared, they cannot be declared unequal. The comparison is made along those dimensions that are considered to be important in the specific context [33]. Other criteria are irrelevant. Which dimensions are important is different for different contexts and objects.

Value Systems Sociologists use the term 'value system' for the system that guides the judgment of what is true, beautiful, good, proper or desirable in human life [24, 33]. Value systems are in general very stable, but they can change due to [34]:

- Generation, although in general only small differences occur in one life itself;
- Changing conditions in the life cycle, e.g. age-composition in society;
- Periodical influences, e.g. incidences of major importance (such as natural disasters).

The social environment—culture, economics and politics—influences the human experience [27]. People experience these influences at different levels.

Societies are divided into a series of domains or levels. The higher levels encompass the lower levels, and the higher levels are considered to have more value [33]. These levels include: micro level: individual, small group, e.g. domestic; meso level: organization, neighborhood, regions, political party, usually more inclusive communities, related to lower values such as honor, power, wealth; and macro level: populations, religions, at an overarching level, that relates to higher values such as ideas about what is ultimately important in life.

People have assumptions about the ends and purposes of human existence, the nature of knowledge, what human beings have the right to expect from each other and the gods, about what constitutes fulfillment and frustration [33].

While the economic view looks very much from the perspective of self-interested calculation, sociologists acknowledge the existence of kindness or adherence to idealistic principles [33].

Culture Value systems differ from one culture to another [24]. Anthropology deals with the way individual personality and social relations are mediated by culture, and how values become ensconced in symbols and meanings [33]. It studies people whose understanding of the world and whose interests and ambitions are most different from our own, aiming to understand collective differences. From this perspective the Western society is the odd one out compared with the rest of the world because of the supreme value it places on individuals (e.g. the immortal soul), and its focus on economic values. Other societies are more holistic [33].

Culture defines the worldview that people have. At an individual level people perceive risks and have concerns, but the socially constructed myths of their culture provide the background for their judgments [17].

The most widely used model for understanding different cultures is the 5D model developed by Geert Hofstede [37], which has been used in extensive cultural studies.

Cultural differences manifest themselves in many ways. From the many terms used to describe manifestations of culture, the following four together cover the total concept rather neatly: symbols, heroes, rituals and values. Symbols represent the most superficial, and values the deepest manifestations of culture, with heroes and rituals in between. The core of culture is formed by values. They deal with: evil versus good; dirty versus clean; dangerous versus safe; forbidden versus permitted; decent versus indecent; moral versus immoral; ugly versus beautiful; unnatural versus natural; abnormal versus normal; paradoxical versus logical; irrational versus rational. Values are acquired early in our lives, mostly in the receptive period of learning up to the age of twelve.

Values are, more than practices, the stable element of culture. In interpreting people's statements about their values, it is important to distinguish between the 'desirable' and the 'desired': how people think the world ought to be versus what people want for themselves. Culture is the unwritten book with the rules of the social game that is passed on to newcomers by its members, nesting itself in their minds. Hofstede identified five dimensions of cultures. A dimension is an aspect of a culture that can be measured relative to other cultures. These dimensions have been used in many studies ever since. The five universal dimensions are listed in Table 3.5.

Table 3.5 The five universal dimensions of culture as defined by Hofstede [37]

Dimension	Explanation
Power distance	The extent to which the less powerful members of institutions and organizations within a country expect and accept that power is distributed unequally.
Collectivism versus individualism	Individualism pertains to societies in which the ties between individuals are loose: everyone is expected to look after himself or herself and his or her immediate family. Collectivism, as its opposite, pertains to societies in which people from birth onward are integrated into strong, cohesive in-groups, which throughout people's lifetimes continue to protect them in exchange for unquestioning loyalty.
Femininity versus masculinity	Society is called masculine when emotional gender roles are clearly distinct: men are supposed to be assertive, tough, and focused on material success, whereas women are supposed to be more modest, tender, and concerned with the quality of life. A society is called feminine when emotional gender roles overlap: both men and women are supposed to be modest, tender and concerned with the quality of life.
Uncertainty avoidance	The extent to which the members of a culture feel threatened by ambiguous or unknown situations.
Long-term versus short-term orientation	Long-term orientation stands for the fostering of virtues toward future rewards—in particular, perseverance and thrift. Its opposite pole, short-term orientation, stands for the fostering of virtues related to the past and present—in particular, respect for tradition, preservation of 'face', and fulfilling social obligations.

All aspects of consumer behavior are culture-bound. People of different countries have different value orientations that cause variations in their preferences for products and brands [19]. Culture also affects the consumers' decision processes: consumers from a culture who value collectivism are more likely to look at their peers before deciding to purchase a certain product or brand than those from a more individualistic culture. Understanding these differences is important when considering innovations with a wide target audience. What works in one culture may not work in another. Cultural differences, when well understood, can also help to spread innovations. In a culture with low uncertainty avoidance an advertisement with humor could be a very good communication strategy to support the market introduction. But this would be totally unacceptable in cultures with high uncertainty avoidance. Here, the best market introduction strategy would be one that provides structured information about the innovation, with more explanation of how it works.

What is interesting is that cultural significance can also change the value of products [56]. A good example is the Trabant cars from the former East Germany. These cars were considered inferior to those from West Germany, and soon became obsolete after the fall of the Berlin Wall in 1989. But after the initial negative sentiments associated with East Germany faded, the Trabant became a collector's item. It represents an era and culture, and therefore has significant cultural value to specific groups of people.

Sentimental and Historical Value Sentimental value relates to the history of a product or its uniqueness [33]. The history of a product, in other words its origin, matters in the perception of value. You can buy fish from two different fishermen, but one may have a better reputation than the other. For the person buying and consuming the fish, the fisherman from whom it is obtained influences the value of the fish. This is not an absolute value but a sentimental value, related to the emotions of a specific person in a specific context. Sentimental value is also common for heirlooms. Pieces of jewelry or furniture that may not have a very high absolute value can acquire a high sentimental value to specific people who appreciate their history.

People may also attach sentimental value to products associated with events that they consider important. Examples of such items are the colorful wristbands that teenagers are given when they visit the clubs on the party island Ibiza. For them they bring back memories of a great holiday, and they associate them with the 'in-crowd' of those who have been there. They reflect a sense of belonging to a community that is special to them.

A product can also have an absolute historical value, such as a genuine piece of stone from the former Berlin Wall. Likewise, pieces of art from famous artists and very unique products such as a first print of a famous book have historical value.

Symbolic Value People can give meanings to objects beyond their use value [56]. These symbolic meanings are attributed to the objects for what they signify to those people. The objects themselves do not contain this value, but the value is created through shared constructs and known symbols [5]. The symbolic value addresses the relationships between people, it sends a social message [33]. An example is the honorary decorations given to people who have done something notable for society, such as war veterans. These decorations, often medals, represent the symbolic value of honor. Similarly, it is not the crown itself that contains the power of a king, but its symbolic value represents that power.

Authenticity is important for symbolic value. When people adopt innovations, it is important that these innovations reflect who they are and aspire to be in relation to how they perceive the world [31]. If there is no match, the symbolic value is lost. A person may buy sports clothes of a brand that is associated with high-sports performance, but if he is not perceived as a good sportsman by others that value is not reflected.

Reciprocity As sociology deals with relationships between people, it is also important to consider both sides. This means that the way people behave toward each other is an important aspect Reciprocity is defined as the practice of exchanging things with others for mutual benefit [58]. Reciprocity can take different forms. It can be open-ended—meaning that no immediate return of value is made, or it can be closed—when a direct return of similar value is made. Open-ended reciprocity is common among close kin: family members do something for you because they know you would do the same for them. In most other cases, reciprocity is more closed and balanced: people feel obliged to deal on a moral basis, as they feel a permanent sense of mutual obligation [33].

In anthropology a lot of research has been done on so-called gift economies, in which valuable goods and services are regularly given without an explicit agreement for immediate or future reward. In the Western societies, altruism—behavior that takes into account the interests of others, as opposed to egoism and selfishness—comes closest to such systems. In most cases, however, giving something comes with getting something, for example access, influence or recognition. Some would argue that true philanthropy rarely exists, and that it is mostly a form of 'narcitropy'—where people crave admiration [31].

Money and Greed Money comes in many forms. Next to conventional coins and notes; credit cards, cheques, vouchers, bank accounts and shares represent money. Moreover, other items like stamps, works of art or diamonds are in a wider notion also recognized as kinds of money [85].

Sociologists and anthropologists have also studied the effects of money and the market mechanism on society. They have concluded that the market has created a kind of alienation from true value. The market connects factories and households, but makes the transactions anonymous. People in the factory are disconnected from the users of their products, and people using the products at home are no longer aware of the effort that has been put into them by workers in the factory. Each is effectively invisible to the other [33].

Money can put an activity in an unfavorable light in two ways. When people are paid for an activity this changes the activity from 'just an activity' into work, which almost always involves a certain amount of dislike. Secondly, if a service (highly personal, using one's own time and skills) is exchanged for money (highly impersonal) people start seeing it as something on which they are losing out because the reward, money, is possibly exploitative [84].

Money is a concrete token of value. Although money in itself carries no specific value, it only has value when it is exchanged for something else. But over time people have lost this notion, and tend to see money as the origin of value. This also means that people have come to see money as an end in itself. They aim to get more of it, as it provides them with the potential to buy things in the future [33]. In this sense, money is not only a tool, but also a drug. Drugs can be very strong motivators; they are often addictive; an attraction to them frequently has bad consequences for the individual; and they give immediate rewards where 'real' motivators can only do so over an extended period [44].

Sociological Value Concepts

Sociology is the study of the development, structure and functioning of human society [64]. Social value is defined as the non-economic value that society puts on a resource and that is recognized by most, if not all, people, such as the benefits to human health of clean air and water [60].

Table 3.6 Value definitions and concepts from a sociological perspective

	User	Organization	Ecosystem	Society
Value definitions	Value as a meaning Signification Symbolic value Sentimental value	Social significance Image value	Network value Affiliation	Cultural value Historical value Social significance
Value concept	Belonging	Social responsibility	Reciprocity	Meaningful life

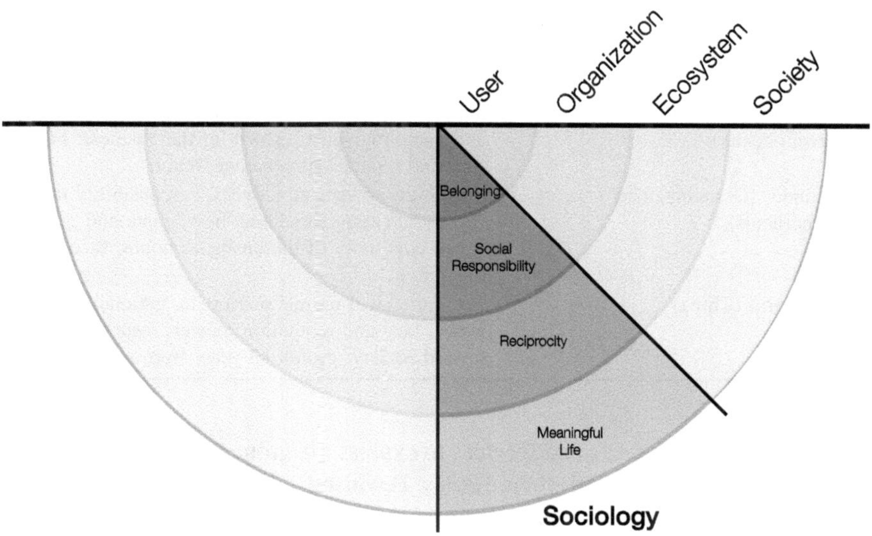

Fig. 3.5 Sociological value concepts in the value framework

Explanations of social order tend to be macro-theories, which focus on society as the unit of analysis, although studies of family obligations, crime and leisure raise issues of social order at the micro level [63]. In general it can be said that sociology focuses on social relationships.

For sociologists it is in relationships that things (products and services) take on value. Table 3.6 shows the relevant definitions as well as the derived value concepts. The sociological value concepts are part of the Value Framework, as shown in Fig. 3.5.

Belonging The sociological perspective on the individual, end-user-level value orientation translates into 'belonging'. Sense of belonging is an important parameter in people's happiness (see also the earlier section on motivational values in the psychological view). Through the group with which people identify, or to which they would like to belong, they have a cultural perspective that provides the context for attribution of a meaning to products and services. This can be related to small groups (such as heirlooms in a family context) as well as larger groups or even whole societies (e.g. religious artifacts or Mao relics).

Table 3.7 Perceived goals of successful business leaders in different cultures [37]

Typical role of business leader	Perceived goals
The executive (United States)	Growth of the business; this year's profits; personal wealth; power; staying within the law; respecting ethical norms
The manager (Britain, New Zealand)	This year's profits; staying within the law; responsibility toward employees; continuity of the business; patriotism and national pride; respecting ethical norms
The family manager (India, Jamaica, Bahamas)	Continuity of the business; family interests; patriotism and national pride; personal wealth; profits 10 years from now; this year's profits
The family entrepreneur (Latin Countries, Australia, Hungary)	Family interests; personal wealth; power; this year's profits; game and gambling spirit; growth of the business
The entrepreneur (Hong Kong, Hawaii-Asian)	Profits 10 years from now; creating something new; game and gambling spirit; growth of the business; honor, face and reputation; personal wealth
The founder (Germany, The Netherlands)	Responsibility toward employees; responsibility toward society; creating something new; game and gambling spirit; continuity of the business; honor, face and reputation
The mandarin (China)	Respecting ethical norms; patriotism, national pride; honor, face and reputation; power; responsibility toward society; profits 10 years from now.

People can use products and services to express the group to which they belong, as the Hells Angels do with their Harley Davidson motorcycles. They can also attach a specific symbolic meaning to products because of a shared history or specific event, such as the 'Make Poverty History' white wristbands. Such items represent a symbolic value to people who relate to a specific context.

For innovations it can be important to understand these mechanisms because they can influence the speed of adoption positively and/or negatively. Apple's iPod is claimed to have become a success because of the high visibility of the typical white earphones: when well-known artists were photographed wearing these, it created a 'me-too' hype among their fans. But it can also work the other way round. Lonsdale was in the news as a brand that was popular among skinheads, and became associated with racism. This is a reason for many people not to want to wear the brand, as they do not want to be associated with such a group.

Social Responsibility Social responsibility is intended to suggest a broad concern with the impact of the organization's behavior on society. The concern is with ultimate outcomes or results, and not simply with policies or intentions. What are considered to be the most important social responsibilities is culture dependent.

Hofstede identified six perceived goals of successful business leaders in seven (sub-)clusters of countries [37], which are shown in Table 3.7. This also means that social responsibility has different connotations in different cultures. However, there is a growing amount of literature that connects social responsibility to

Table 3.8 Value definitions and concepts from an ecological perspective

	User	Organization	Ecosystem	Society
Value definitions	Human development Creativity Spirituality	Business Spirituality Creativity	Earth centered values	Biodiversity Gross National Happiness
Value concept	Eco-footprint	Eco-effectiveness	Sustainability	Livability of the environment

sustainability [68]. This specific aspect will be addressed in the following section on the ecological perspective.As the list in Table 3.7 shows, there are significant differences in what is expected from companies in certain cultures. Where one culture focuses on short-term economic and more personal goals, others are looking for longer term goals that benefit larger groups of people. When launching innovations it is important to understand these differences, and to take into account what they mean for innovations to become successful. It is not necessary to abandon goals that are not listed in a certain culture, but those that are considered important have to be included in the proposition to make an innovation a success in that specific culture.

Reciprocity The sociological perspective at the ecosystem level translates into 'reciprocity'. If the members of the ecosystem are to have fruitful relationships with each other in the longer term, it is important that they all contribute to and benefit from the system. Not all relationships in the system need to be directly reciprocal: they can be indirect, such as through a third party, or reciprocity may occur over time. Nevertheless long-term stability of the relationships in the ecosystem means there needs to be a balance for all members. They are mutually interdependent.

All the members have to build on their own competences and stay authentic in their offerings because only then can they make their maximum contribution and support the long-term strength of the ecosystem. Gilmore & Pine emphasize that each organization should stick to its own reason of being. This means that only companies should be in the business of commerce—buying and selling in the pursuit of personal and corporate profit. Other organizations such as educational, governmental, charitable and church-based institutions have other motives, and each has an appropriate method of funding, see Table 3.8 [31]. Businesses are supposed to make a profit for the sake of their stakeholders and employees, hence payment and fees from customers are an appropriate method of funding their activities. Government subsidies are perceived by an inauthentic method of funding for for-profit organizations. Charity organizations rather accept money from individuals, as payments received from government agencies or business owners always come with strings attached. For educational institutes that aim to advance knowledge of individuals and mankind, the appropriate method of funding is grants and tuitions to cover the costs of operating the school.

Reciprocity means that in the end all members of the system contribute from their own strengths and competences, and are paid in return in a value that is of worth to them. For commercial organizations this payment will usually be in the form of money, but for other organizations it can be in different types of value—a business could contribute to a school by providing guest lectures by its experienced managers.

The concept of reciprocity needs to take into account that different types of values, as parties in the system, can be of a different nature. In a reciprocal relationship in an ecosystem, different types of values can be exchanged that together create the whole. Businesses and government organizations will need to work together on innovations that alleviate societal issues. This is not a problem as long as each organization sticks to its own purpose, and plays a role that fits with its purpose, competences and methods of funding.

Meaningful Life Value at the level of society from a sociological perspective translates into living a meaningful life. The ultimate value of society is the greatest happiness of the greatest number of people [93]. This means that the sense of belonging at the individual level translates into contributing meaningfully to society. Well-being increases if as many people as possible contribute from their own strengths in a meaningful way to a bigger cause. There is a strong connection between the psychological and sociological perspectives. Society provides the bigger picture, and the reference to which the individual can contribute.

The Ecological View

The word ecology was first used in 1866 by Ernst Haeckel, who aimed to find a name for a branch of science that studies phenomena in the physical and biological environment of organisms that affect their survival in the broadest sense. He was heavily inspired by Charles Darwin's theory of evolution. But it took until around 1920–1930 before the first integrative perspectives emerged that looked into new concepts like 'biocenosis', the integrated community consisting of all living beings associated with a given habitat or set of environmental conditions; the 'biosphere', looking into biogeochemical cycles; and the 'ecosystem', introduced by the British botanist Arthur G. Tansley in 1935 [57].

Ecology also refers to the social movements for environmental protection that emerged in the 1960s. At that time there were growing concerns about issues like chemical pesticides and nuclear power. In the 1980s problems with acid rain contributed to further worries, as did ozone depletion [11], deforestation and the extinction of species in wild life in the 1990s. The Ecology Movement aims to protect against environmental destruction, and holds the view that nature is a value in itself. Even today there is still much disagreement about the human role in the preservation of biodiversity and the maintenance of unspoilt natural reserves. Some hold the view that we should stop regarding ourselves as superior to other organisms and respect the living environment as a whole, with everything having the same right to live and flourish. Others simply feel the need to take action to preserve animal and plant life from extinction.

Value in the perspective of ecology considers the earth as a whole, with man just being part of a larger ecosystem. This creates a more holistic view of values than the other social sciences as discussed before.

Earth-Centered Values We misconceive our role as humans in the functioning of the earth if we consider that our historical mission is to 'civilize' or to 'domesticate' the planet. Humans are not here to control. We are here as an integral part of a larger earth community. We need to appreciate the spontaneities found in every form of existence in the natural world: the wild component of nature, that is uncontrolled by human dominance. We should consider ourselves as part of a bigger system, and move from a human-centered to an earth-centered norm of reality and value [4].

Many have raised the issue that with our industrial civilization we have plundered the earth [23]. Natural resources are being depleted at an unsustainable rate. It might have taken around two billion years for fossil fuels to accumulate, and experts disagree on when we will have guzzled them all [32]. But we are certainly consuming them at a much higher rate than their creation [46]. It will be necessary to take the larger system into account in deciding on the use of scarce natural resources, and to take an earth-centered view, instead of just looking from our own human perspective.

As Lester Brown puts it: "We have been hiding some of the costs off the books." And quoting Oystein Dahle, former vice president of Exxon for Norway and the North Sea: "Socialism collapsed because it did not allow the market to tell the economic truth. Capitalism may collapse because it does not allow the market to tell the ecological truth." [10].

One which is not priced is not valued, and if our open lands, clean air, clean water and healthy forests are not valued, the earth, when it is this flat and this crowded, will become a very hot, no-cost landfill very fast. Markets have failed by underpricing goods and service by excluding the price of their externalities, and the impact of that underpricing has highly negative economic, health and national security implications [28].

Value of Biodiversity Humans depend to a large extend on wildlife. These services of the ecosystem include maintaining the fertility of soil, cleaning of water, regulating the climate and controlling pests [16]. For example, an entire population of a given tree species might provide the global service of carbon sequestration, and a regional population of the same tree species might provide a water filtration service that benefits local communities [45].

The value of biodiversity can be classified into six major categories [25]:

- Direct, extractive, use value: production of timber or the collection of animals and plants for food;
- Direct, non-extractive use value: recreation and tourism;

- Ecological function value: flood control, climate regulation, photosynthesis, nutrient cycles, waste assimilation;
- Option value: the potential use of plants for future drugs, genes for plant breeding, importance of plants as a source of chemical substances to complement new technologies, or as a substitute for depleted resources;
- Existence (non-use) value: sympathy for the natural environment and the mere existence of particular species;
- Bequest value: altruistic benefit from the knowledge that others might benefit from a resource in the future (e.g. younger generations).

Conservation of biodiversity is therefore important. Economic valuation techniques can to a certain extent be applied to calculate the direct and indirect value of biodiversity and ecosystems, but it remains difficult to express these in terms of hard numbers. Nevertheless, a consequence of not valuating biodiversity in economic models is that many developing countries suffer devastating deterioration of their natural resources. One example is the tropical forests that are burnt to create farmland for relatively unproductive use in Indonesia, Malaysia or Costa Rica.

All conservation is local—but we are globally responsible, warns Milton Friedman. "The palm oil that fried up your French fries today may have come from a chopped-down tropical forest in Indonesia, which in turn helped to contribute to climate change that is intensifying the drought in your backyard. That is why deforestation is everyone's problem." [28].

The Value of Nature The importance of natural surroundings for human well-being and health has been discussed in many studies [47]. Nature has a proven restorative capacity to alleviate stress that people experience, but natural environments can also prevent stress through aiding in the recovery of this essential resource [40].

Contact with nature has proven to promote health and well-being. The results of various scientific studies have proven that [47]:

- There are some known beneficial physiological effects that occur when humans encounter, observe or otherwise positively interact with animals, plants, landscapes or wilderness;
- Natural environments help recovery from mental fatigue and are restorative;
- There are established methods of nature-based therapy (including wilderness, horticultural and animal-assisted therapy) that have been successful in healing patients who formerly did not respond to treatment;
- When given a choice, people prefer natural environments (particularly those with water features, large old trees, intact vegetation or minimal human influence) to urban spaces, regardless of nationality or culture;
- The majority of places that people consider favorite or restorative are natural places, and being in those places has a recuperative effect;
- People have a more positive outlook on life and higher life satisfaction when they are close to nature (particularly in urban areas);

- Exposure to the natural environment enhances the ability to cope with and recover from stress, to cope with subsequent stress and to recover from illness and injury;
- Observing nature can restore concentration and improve productivity;
- Having nature close at hand, or just knowing it exists, is important to people regardless of whether they are regular 'users' of it.

Even a view from a window can have a positive impact on well-being [39].

Human Development The Human Development Index is a compound indicator of economic development established by the United Nations, which aims to get away from purely monetary measures by combining per capita Gross Domestic Product with life expectance and literacy in a weighted average. Human Development is a development paradigm that is much more than the rise and fall of national incomes. It is about creating an environment in which people can develop their full potential and lead productive, creative lives that are in accord with their needs and interests. People are the real wealth of nations. Development is therefore about expanding the choices that people have to lead their lives [61, 89].

Putting the economy back in the service of people and life, and not the other way round, has been the main concern of the Human Scale Development approach. This is in line with Aristotle's conceptualization that the 'oikonomy' has to deal not just with living, but more fundamentally with the 'art of living well'. Within this paradigm it was suggested that the best development process will be the one that enables improvement in people's quality of life, allowing people and communities to be coherent within themselves. A crucial aspect in the concept of human scale development is the idea that people have multiple and interdependent needs that interact and interrelate in a systemic way, and there is no two-way correspondence between needs and satisfiers [49]. A satisfier can contribute simultaneously to the satisfaction of various needs; or conversely a need can require various satisfiers to be met. Satisfiers can be of five different types [14]:

- Synergetic satisfier: those which, by they way in which they satisfy a given need, promote and contribute to the simultaneous satisfaction of other needs;
- Singular satisfier: aiming at the satisfaction of targeted single needs, this kind of satisfier is neutral regarding the satisfaction of other needs;
- Destructive satisfier: elements of paradoxical effect. Applied under the pretext of satisfying a given need, they not only annihilate the possibility of its satisfaction, but they also render the adequate satisfaction of other needs impossible;
- Inhibiting satisfier: those which by the way in which they satisfy (generally over-satisfy) a given need, seriously impair the possibility of satisfying other needs;
- Pseudo-satisfier: elements that stimulate a false sensation of satisfying a given need. Although they lack the aggressiveness of destructive satisfiers, they may on occasion, in the medium or long term, annul the possibility of satisfying the need they were originally aimed at.

In participatory processes with people, individuals and communities, it is important that people are self-aware about their preferences. Comparing the new, proposed situation with the current, actual situation reveals if a certain proposition will indeed improve the quality of life of people in the important aspects. In-depth discussions with various stakeholders are needed to reflect on the proposal, and to improve it further to better meet the different needs [14].

Environments Stimulating Creativity The current economy is increasingly powered by human creativity. Creativity, the ability to create meaningful new forms, is the decisive source of competitive advantage. Places that harness human creativity become important, and as Richard Florida emphasizes, such places are not necessarily bound to traditional areas like R&D departments in companies. It is much more likely that, with the merging of private and working life, hotspots of creative activity will emerge that attract people to live and work in settings in which creativity flourishes. His model builds on 3Ts for growth: Technology—measured by innovation and high-tech industry concentration; Talent—not 'human capital' as it is usually measured in terms of numbers of people holding higher education credentials, but creative capital, which is talent measured functionally by the numbers of people actually in creative occupations; and Tolerance—places that are open and tolerant have an edge in attracting different kinds of people and generating new ideas [26].

The group of 'cultural creatives', which covers 24% of adults in the US and most likely similar percentages in Western Europe, are altruistic and often less concerned with success or making a lot of money, although most live comfortably with middle- to upper-class incomes. They value ecological sustainability, globalism, women's issues, social conscience, self-actualization and spirituality [72].

Innovation hotspots have created places where public and private players collaborate to create new solutions. Finland, ranked number 2 on the list of worldwide innovation hotspots, has created an innovation system in which investment in public education, and a flexible funding scheme for public and private R&D and innovation activities, have resulted in novel, commercially significant ideas [38].

Spirituality Related to the focus on humans is the increasing attention to living a meaningful life, not only as a private person, but also as workers in the knowledge age. As the best-selling author Stephen Covey phrases it: "The crucial challenge is to find your inner voice, and inspire others to find theirs" [13]. It may be related to the fact that more and more people in Western societies are able to trust that the lower needs in Maslow's hierarchy are being met. Workers are increasingly and naturally beginning to explore their higher-order needs, seeking more than money from work, looking for social esteem and self-actualization [87].

Spirituality is not linked to religious notions. It is defined as the search for connectedness and meaning. Spirituality is an innate and universal search for transcendental meaning in life. It involves the desire to do purposeful work that serves others and to be part of a principled community. It involves the yearning for connectedness and wholeness that can only be manifested when one is allowed to integrate his or her inner life with one's professional role in the service of a greater good [2].

The business world focuses on performance and innovation exclusively through rational understanding. This view overlooks the reality of people, the very resource for the spiritual power of innovation. Inspiration is an important capacity for great business performance, but it cannot be fully understood through rational concepts. Paul de Blot, professor of Business Spirituality at Nyenrode Business University, aims to understand the 'soul' of doing business. He proposes organizational spirituality as a system at three levels: the level of doing—the beginning level of physical energy; the level of interaction—which takes this doing to a greater depth; and the level of being—which reaches the roots of spiritual energy and well-being. From this last level, the spiritual energy connects the other levels and creates perfect harmony among the physical, mental and spiritual forces. Strong business spirituality is a powerful harmonizing force for the individuals who make up families and organizations. In contrast, disharmony is destructive at all levels [18].

To respond to the spiritual needs of employees and allow them to be successful, leaders and managers must embrace the principles of the business spirituality paradigm. They could operate under the assumptions and values of consciousness, collaboration, inner orientation and the like. They should lead 'from the heart'. An organization should be viewed as the cooperation between people as spiritual beings. Insight into the deeper meaning and drives of people can inspire creative renewal [9].

Gross National Happiness (GNH) The concept of Gross National Happiness was already mentioned in the section on economic value, as an attempt to define an indicator that measures quality of life in a more holistic way than its economic counterpart GDP. The term was coined by Bhutan's former King Jigme Singye Wangchuck in 1972 [54]. The aim was to find a measure to support the development of Bhutan's economy in line with Buddhist spiritual values. To enable policies and decisions to be in line with the concept, a GNH index was developed that includes nine dimensions: psychological well–being; time use (whether respondents feel they have sufficient time for various non-work activities); community vitality (strengths and weaknesses of relationships and interactions within communities); culture (diversity and resilience of cultural traditions); health; education, environment (perceptions and ecological knowledge); living standards; and governance (perceptions of equity, honesty and quality). These dimensions are equally weighted because each dimension is considered to be relatively equal in terms of its intrinsic importance as a component of Gross National Happiness [3].

The concept of GNH is interesting as it tries to make a holistic view tangible and measurable. And although critics remark that it is hard to establish accurate numbers for it, it is useful in establishing progress and highlighting dimensions in which shortfalls from sufficiency are most severe [6].

Ecological Value Concepts

The ecological perspective on value is the most comprehensive of the social sciences. It covers not only the social relationships of people, but also their

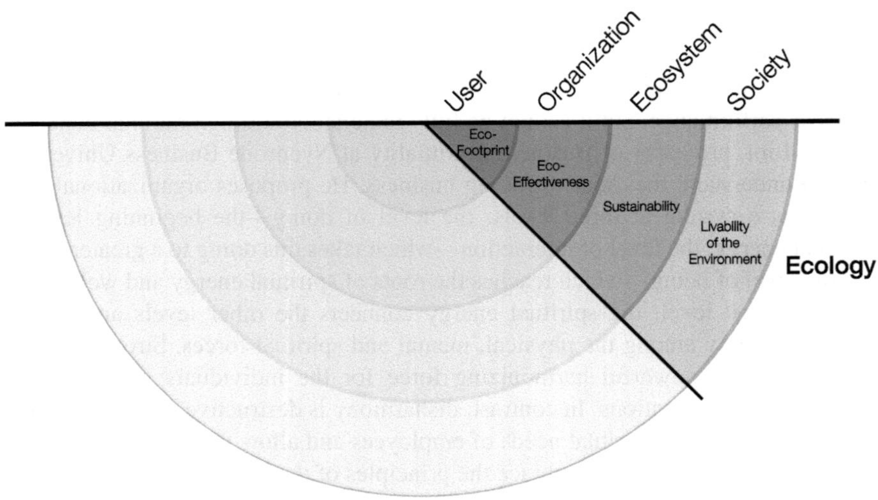

Fig. 3.6 Ecological value concepts in the value framework

relationships with their physical surroundings. As the social relationships have already been covered in the previous section, the values, definitions and concepts highlighted in this section focus on the relationships with the physical environment. This may not justify the holistic view that fits the ecological perspective, but it helps to avoid too much overlap with the sociological perspective. The four value concepts that fit this focus on the physical environment for each of the value levels are shown in Table 3.8 and Fig. 3.6, and are described in more detail in the sections below.

Eco-Footprint The environmental value at a user level can be defined by the individual's ecological footprint. Each person requires a certain amount of biologically productive land and sea area to generate the resources he or she consumes and to absorb or render harmless the resulting waste. By selecting products and services, users have the options of choosing between different alternatives with different ecological footprints. This footprint is related not only to the production process, but also to the environmental impact during use and to the disposal of the product as waste. One product might contain more hazardous materials (such as cadmium), but on the other hand its energy consumption during use may be lower.

The concept of measuring ecological footprints was originally intended to measure and track a national economy's energy and resource throughput, and translate it into the biologically productive areas needed to produce those flows. This framework allows human consumption to be compared with natural capital production at the global and national levels [94]. An analysis of the ecological footprints of cities reveals that cities cause accelerated global ecological decline and are not sustainable by themselves. At the same time, their inhabitants can play a major role in helping to achieve global sustainability [73]. So far, estimates are that with current consumption patterns of the Western nations we will need at least three times the available land on planet earth to sustain us.

However, it looks as if it is not so easy for individuals to behave in an eco-friendly way [30]. There is a lot of confusing information, and it is difficult to judge what is best. David MacKay made a tremendous effort to create a simple, comprehensible and fact-based method that should help individuals to make conscious choices, and help to drive their eco-footprints down to sustainable levels. It enables the creation of a personal balance sheet to check whether personal energy consumption can be matched with conceivable sustainable production [46]. Understanding the energy consumption of products that we use on a daily basis, like food, transportation and household appliances gives people a much clearer insight into their total eco-footprint. But awareness is only the first step to changing behavior.

Innovations should always consider the total eco-footprint they generate (which will be discussed in more detail in the next section), but transformational innovations could even contribute positively to reducing users' average footprints.

Eco-Effectiveness At an organizational level, the ecological value concept is related to eco-friendly innovation. Eco-design has already been practiced for about 20 years. Although some companies are front-runners, there is still not enough dissemination of knowledge and education to ensure that more companies adopt the approach, and go beyond incremental improvements [88]. Tools have helped to provide insight into the total ecological footprint of a product, from development and production through to use and disposal. Examples of such tools such as Life Cycle Assessment and Environmental Effect Analysis are described in the abundantly available literature, both scientific and more popular.

The importance to organizations of embarking on the 'green' journey is not just to avoid the risk of being labeled as 'polluters' and gaining the vicious attention of environmental action groups. 'Out-greening', as Milton Friedman calls it, will become a strategy for achieving competitive advantage in a variety of fields. Today, whatever kind of product you make or service you offer, it can quickly and easily be copied and sold by anyone anywhere. But 'how' you do your business, 'how' you keep your promises and 'how' you relate to customers, colleagues, suppliers and the communities in which you operate are much more difficult to copy if you are doing them well. This creates opportunities for sustainable differentiation; opportunities to 'out-behave' the competition [28].

In their book 'Cradle-to-Cradle', McDonough and Braungart introduced the approach of eliminating the concept of waste altogether. As recycling inherently becomes 'down-cycling' because material properties are downgraded, cradle-to-cradle calls for a different approach that creates much more elegant environmental solutions [50]. Cradle-to-cradle design is presented as an alternative approach to the strategies of 'zero-emission' and 'eco-efficiency'. Eco-efficiency seeks to reduce the unintended negative consequences of processes of production and consumption. But eco-effectiveness, in contrast, with as its central component cradle-to-cradle design, is a positive agenda for the conception and production of goods and services that incorporate social, economic and environmental benefits, enabling triple top line growth [7].

The eco-effective innovations that are the aim of cradle-to-cradle design fit very well with the view of transformational innovations. Examples include car tires designed to attract and capture harmful particles, thereby cleaning the air instead of polluting it. Many more innovations can be thought of that make positive contributions to the overall quality of life.

Sustainability Ecology at eco-system level deals with sustainability. As Milton Friedman clearly indicates: "Acting with sustainable values cannot be reduced to a simple checklist. Sustainability is about the disposition, the mindset, and behaviors which shape and sustain relationships—relationships with family, friends, customers, investors, employees, borrowers, fellow citizens, the community, the environment, and with nature. In both the natural world and the financial world it means that you think and then behave in a way that literally sustains—sustains the world around you, sustains business relationships, sustains personal relationships, sustains your community, sustains your country, sustains the planet, and sustains your relationships with your grandchildren and with generations to come. And that mindset automatically leads you to the values that connect us deeply as people to other people and as people to institutions, to communities, and to the environment—values such as transparency, integrity, honesty and shared responsibility. This mindset also always leads you to think about the impact your actions will have over the longer term. Sustainability is not a euphemism for charity work or socialism; it is an essential ingredient for winning. Practicing sustainability not only helps a company to survive, it helps it thrive over the longer term. Sustainable values do double duty. They preserve, restore, and enhance companies, while simultaneously generating renewable resources, innovation, advantage and prosperity. There is real value in sustainable values." [28].

In the biological context, sustaining an ecosystem's ecological integrity has been defined as one of the most important goals for its management. It encompasses the maintenance of viable populations of all native species, and the representation of all native ecosystem types in their natural range of variations. Sustainability must be a primary objective. In practice, consensus must be sought among the stakeholders on the use of the resources and processes. It also requires long-term management dealing with timescales beyond the fiscal-year scope of most organizations. Successful ecosystem management also requires making the system adaptable to change [12].

Livability of the Environment At a societal level, ecological value can be captured in the value concept of livability of the environment. The livability of the environment and the importance of natural surroundings for human well-being and health are one reason. But the physical beauty of nature is certainly also important. It is hard not to appreciate the contours of a mountain landscape, the ambient colors of the setting sun, the power of a lion, the quietly floating clouds or the song of a bird. Studies have revealed that even unspectacular or below-par natural views elicit higher preference than urban views [42].

Preserving nature to increase the livability of the environment has to be part of innovation. Preserving nature can also be an inspiration for transformational

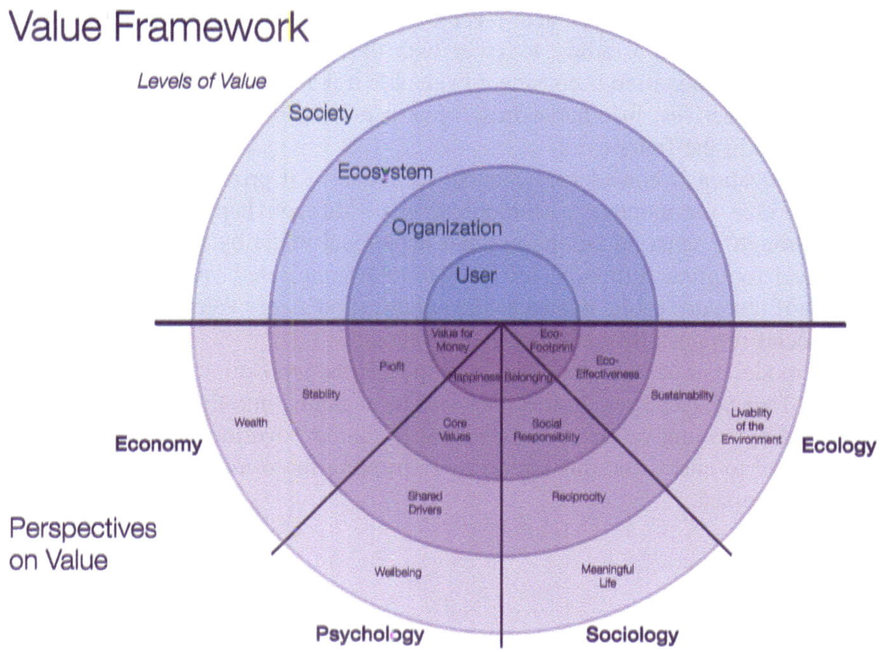

Fig. 3.7 The value framework: an integrated view on value from social sciences

innovation. An example of this is 'Metamorphosis', a design probe created by Philips Design [65]. It explores how we have become separated from the natural world, both in terms of our surroundings and how we perceive and manage our time. It reconnects the spaces in which we live and work to the outside world, where seasons and time awaken powerful instinctive responses in us. The concepts in 'Metamorphosis' channel natural qualities and form a barrier against the unnatural elements we are bombarded with. They respond both to the condition of atmosphere, light and sound outside as well as to the people who share the spaces with them. Rather than creating 'artificial' air, sound or light, they transport the natural external world into the places we live, work and recover.

The Value Framework

The combination of the different views of value from the relevant social sciences, with the four levels of the previous chapter provides a comprehensive integrated view on value, as shown in Fig. 3.2.

The picture shows the four levels as concentric circles, each higher level encompassing the lower ones Fig. 3.7.

In the bottom half of the picture the foundation is laid for value creation. For each of the levels and each of the perspectives, the overarching value concept is given. Each of the social sciences—economics, psychology, sociology and ecology—is indicated on a radius, together with the overarching value concepts for each level. An innovation is considered valuable if it addresses the four levels from all four perspectives; in other words, when a positive check is made for all the items stated in the framework.

An innovation is considered valuable for a user if it provides economic value through value for money; psychological value through happiness; sociological value through a sense of belonging; and ecological value by reducing the user's ecological footprint. Similarly, an innovation is considered valuable to an organization if it creates profit, enhances the organization's core values and contributes to its social responsibility and eco-effectiveness.

The model can be used to evaluate proposals for innovations at all four levels from all four perspectives. In this way insight is gained into the elements that are addressed by an innovation, and the elements that are overlooked. These insights can be used to enrich the innovation by improving those aspects that are insufficiently addressed.

References

1. Allardt, Erik. 1976. A welfare model for selecting indicators of national development. *Policy Sciences* 4(1): 1973.
2. Ashar, Hanna, and Maureen Lane-Maher. 2004. Success and spirituality in the new business paradigm. *Journal of Management Inquiry* 13(3): 249–260.
3. Bates, Winston. 2009. Gross National Happiness. Asian-Pacific Economic Literature, 23(2).
4. Berry, Thomas. 1999. *The great work: Our way into the future*. New York: Bell Tower.
5. Boztepe, Suzan. 2007. User value: Competing theories and models. *International Journal of Design* 1(2): 55–63.
6. Braun, Alejandro Adler. 2009. Gross National Happiness in Bhutan: a Living Example of an Alternative Approach to Progress. Wharton International Research Experience. September.
7. Braungart, Michael, William McDonough, and Andrew Bollinger. 2007. Cradle-to-Cradle design: Creating healthy emissions—a strategy for eco-effective product and system design. *Journal of Cleaner Production* 15(13/14): 1337–1348.
8. Brock, D. 1993. Quality of life in health care and medical ethics. In *The quality of life*, eds. M. Nussbaum, A. Sen, 95–132. Oxford: Clarendon Press.
9. Broekstra, Gerrit, Paul De Blot SJ, Ruud Heijblom, Jan Bommerez, and Kees van Zijtveld. 2006. De Brug naar Business Spiritualiteit. Radicale Vernieuwing Begint bij de Wortels. Uitgeverij Nieuwe Dimensies, Eemnes, The Netherlands.
10. Brown, Lester R. 2008. *Plan B 3.0 mobilizing to save civilization. earth policy institute*. London: W.W. Norton & Company Inc.
11. Brunnee, J. 1988. *Acid rain and ozone layer depletion: International law and regulation*. Dobbs Ferry, New York.
12. Christensen, Norman L., Ann M. Bartuska, James H. Brown, Stephen Carpenter, Carla D'Antonio, Rober Francis, Jerry F. Franklin, James A. MacMahon, Redd F. Noss, David J. Parsons, Charles H. Peterson, Monica G. Turner, and Robert G. Woodmansee. 1996. The report of the ecological society of America committee on the scientific basis for ecosystem management. *Ecological Applications* 6(3): 665–691.
13. Covey, Stephen R. 2004. *The 8th Habit*. New York: The Free Press.

14. Cruz, Ivonne, Andri Stahel, and Manfred Max-Neef. 2009. Towards a systemic development approach: building on the human scale development paradigm. *Ecological Economics* 68: 2021–2030.
15. Csikszentmihaly, Mihaly. 1990. *Flow: The psychology of optimal experience*. New York: Harper and Row.
16. Daily, Gretchen C., Tore Soderqvist, Sara Aniyar, Kenneth Arrow, Partha Dasgupta, Paul R. Ehrlich, Carl Folke, AnnMari Jansson, Bengt-Owe Jansson, Nils Kautsky, Simon Levin, Jane Lubchenco, Karl-Goran Maler, David Simpson, David Starrett, David Tilman, and Brian Walker. 2000. The value of nature and the nature of value. *Science* 289: 395–396.
17. Dake, Karl. 1992. Myths of nature: culture and the social construction of risk. *Journal of Social Issues* 48(4): 21–37.
18. De Blot, Paul. 2010. Innovation of management in a crisis: The spiritual power of Organisations. in *Spirituality and Business*. eds. S.S. Nandram and M.E. Borden. Berlin: Springer-Verlag
19. Marieke., De Mooij. 2004. *Consumer behavior and culture: consequences for global marketing and advertising*. London: Sage Publications.
20. Diener, Ed, Richard E. Lucas, and Christie Napa Scollon. 2006. Beyond the hedonic treadmill: Revising the adaptation theory of well-being. *American Psychologist* 61(4): 305–314.
21. Diener, Ed, Shigehiro Oishi, and Richard E. Lucas. 2003. Personality, culture and subjective well-being: emotional and cognitive evaluations of life. *Annual Review of Psychology* 54: 403–425.
22. Diener, Ed, and Eunkook Suh. 1997. Measuring quality of life: economic, social, and subjective indicators. *Social Indicators Research* 40: 189–216.
23. Douthwaite, Richard. 1995. *The growth illusion: How economic growth has enriched the few, impoverished the many, and endangered the planet*. Dublin: Lilliput Press.
24. Dumont, Louis. 1980. *Homo hierarchicus: The caste system and its implications*. Chicago: University of Chicago Press.
25. Edwards, Peter J., and Cyrus Abivardi. 1998. The value of biodiversity: Where ecology and economy blend. *Biological Conservation* 83: 3239–3246.
26. Florida, Richard. 2002. *The rise of the creative class: And how it's transforming work, leisure, community, and everyday life*. New York: Persues Books.
27. Forlizzi, Jodi. 2008. The product ecology: Understanding social product use and supporting design culture. *International Journal of Design* 2(01): 11–20.
28. Friedman, Thomas L. 2009. *Hot, flat and crowded: Why the world needs a green revolution—and how we can renew our global future*. London: Penguin Books.
29. Friedman, Thomas. 2005. *The World is Flat*. London: A brief history of the twenty-first century. penguin books.
30. Gatersleben, Birgitta, Linda Steg, and Charles Vlek. 2002. Measurements and determinants of environmentally significant consumer behavior. *Environment and Behavior* 34(3): 335–362.
31. Gilmore, James H., and B. Joseph Pine II. 2007. *Authenticity: What consumers really want*. Boston, Massachusetts: Harvard Business Press.
32. Goodstein, D. 2004. *Out of Gas*. New York: W.W. Norton & Company.
33. Graeber, David. 2001. *Toward an anthropological theory of value: The false coin of our own dreams*. New York: Palgrave Macmillan.
34. Gundelach, p. 1992. Recent value changes in Western Europe. *Futures* 24: 301–318.
35. Helgeson, James G., and Michael L. Ursic. 1992. Information load, cost/benefit assessment and decision strategy variability. *Journal of the Academy of Marketing Science* 21: 13–20.
36. Heskett, John. 2008. Creating economic value by design. *International Journal of Design* 3(1): 71–84.
37. Hofstede, Geert H., and Gert Jan Hofstede. 2005. *Cultures and organisations: Software of the mind*. New York: McGraw-Hill.

38. Kao, John. 2009. "Tapping into the World's Innovation Hotspots." Harvard Business Review, March.
39. Kaplan, Rachel. 1993. The role of nature in the workplace. *Landscape and Urban Planning* 26(1–4): 193–201.
40. Kaplan, Stephen. 1995. The restorative benefits of nature: toward an integrative framework. *Journal of Environmental Psychology* 15: 169–182.
41. Keinonen, Turkka. 2010. Protect and appreciate–notes on the justification of user-centered design. *International Journal of Design* 4(1): 17–27.
42. Kellert, Stephen R. 1993. "The Biological Basis for Human Values of Nature." Chapter in The Biophilia Hypothesis.eds.Stephen R. Kellert, Edward O. Wilson. Island Press.
43. Kluckhohn, Clyde. 1951. "Values and Value-Orientations in the Theory of Action: An Exploration in Definition and Classification." Chapter in Toward a General Theory of Action. T. Parsons and E. Shils eds. Harvard University Press,388–433.
44. Lea, Stephen E.G., and Paul Webley. 2006. Money as tool, money as drug: the biological psychology of a strong incentive. *Behavioral and Brain Sciences* 29: 161–209.
45. Luck, Gary W., Gretchen C. Daily, and Paul R. Ehrlich. 2003. "Population Diversity and Ecosystem Services." Trends in Ecology and Evolution, 18:(7).
46. MacKay, David J.C. 2009. *Sustainable Energy—Without the Hot Air.* Cambridge, UK: UIT Cambridge Ltd.
47. Maller, Cecily, Mardie Townsend, Anity Pryor, Peter Brown, and Lawrence St Leger. 2005. Healthy nature healthy people: 'contact with nature' as an upstrea health promotion intervention for populations. *Health Promotion International* 21(1): 45–54.
48. Marx, Karl. 1993. Capital: A Critique of Political Economy (Volume 1). Penguin Classics.
49. Max-Neef, Manfred. 1992. "Development and Human Needs." Chapter in Real-Life Economics: Understanding Wealth Creation. P. Ekins, and M. Max-Neef eds. Taylor & Francis Ltd, London, pp.197–214.
50. McDonough, William, and Michael Braungart. 2002. Cradle-to-Cradle. Remaking the Way We Make Things. North Point Press.
51. McKinsey. 2000. "Delivering Value to Customers." McKinsey Quarterly.
52. Moore, James F. 1996. *The death of competition leadership and strategy in the age of business ecosystems.* New York: Harper Business Publishers.
53. Mundt, JoNel. 1993. Externalities: Uncalculated outcomes of exchange. *Journal of Macromarketing* 13(2): 46–53.
54. Official Bhutanese Gross National Happiness Website, Centre for Bhutan Studies. http://www.grossnationalhappiness.com/ Accessed 8 December 2010.
55. Oishi, S., E. Diener, E. Sub, and R.E. Lucas. 1999. Values as a moderator in subjective well-being. *Journal of Personality* 67:157–184.
56. Owens, Bruce McCoy. 1999. Unruly readings: Neofetishes, paradoxical singularities, and the violence of authentic value. *Ethnos* 64(2): 249–262.
57. Oxford Companion to the History of Modern Science. J.L. Heilbron, ed. Oxford University Press 2003. Oxford Reference Online. Accessed 3 January 2011.
58. Oxford Dictionary of English (revised edition). ed. Catherine Soanes and Angus Stevenson. Oxford University Press, 2005. Oxford Reference Online. Accessed 7 December 2010.
59. Oxford Dictionary of English Etymology. ed. T.F. Hoad. Oxford University Press, 1996. Oxford Reference Online. Accessed 29 December 2010.
60. Oxford Dictionary of Environment and Conservation. Chris Park. Oxford University Press, 2007. Oxford Reference Online. Accessed 20 December 2010.
61. Oxford Dictionary of Geography. Susan Mayhew. Oxford University Press 2009. Oxford Reference Online. Accessed 28 December 2010.
62. Oxford Dictionary of Human Resources Management.
63. Oxford Dictionary of Sociology. John Scott and Gordon Marshall. Oxford University Press 2009. Oxford Reference Online. Accessed 8 December 2010.
64. Oxford Dictionary of the Social Sciences. Craig Calhoun, ed. Oxford University Press 2002. Oxford Reference Online. Accessed 7 December 2010.

65. Philips Design. Metamorphosis., Design Probe, June 2010. http://www.design.philips.com/shared/assets/design_assets/pdf/news/Metamorphosis_background.pdf. Accessed 5 January 2011.
66. Pine, B.Joseph, and James H. Gilmore. 1999. *The experience economy: work is theatre & every business a stage*. Boston, MA: Harvard Business Press.
67. Pitelis, Christos N. 2009. The co-evolution of organizational value capture, value creation and sustainable advantage. *Organization Studies* 30(10): 1115–1139.
68. Pohl, Manfred, and Nick Tolhurst. 2010. *Responsible business: how to manage a csr strategy successfully*. London: Wiley.
69. Porter, Michael E. 1985. *Competitive advantage*. New York: The Free Press.
70. Prahalad, C.K. 2004. *The fortune at the bottom of the pyramid eradicating poverty through profits*. Englewood Cliffs NJ: Prentice Hall.
71. Ramírez, R. 1999. Value co-production: intellectual origins and implications for practice and research. *Strategic Management Journal* 20: 49–65.
72. Ray, P.H. 1997. The emerging culture. *American Demographics* 19(2): 29–34.
73. Rees, William, and Mathis Wackernagel. "Urban Ecological Footprints: Why Cities Cannot be Sustainable - and Why They are a Key to Sustainability." Chapter in Urban Ecology. John M. Marzluff, Eric Shulenberger, Wilfried Endlicher, Marina Alberti, Gordon Bradley, Clare Ryan, Ute Simon, and Craig ZumBrunnen eds, Springer US, 2008.
74. Robbins, Lionel. 1932. *An essay on the nature and significance of economic science*. London: MacMillan.
75. Rokeach, Milton. 1979. *Understanding human values. individual and societal*. NY: Free Press.
76. Sagiv, Lilach, and Shalom H. Schwartz. 2000. Value priorities and subjective well-being: direct relations and congruity effects. *European Journal of Social Psychology* 30(2): 177–198.
77. Sawhney, Mohanbir S. 1998 "Leveraged high-variety strategies: From portfolio thinking to platform thinking." Journal of the Academy of Marketing Science, 26(1) Winter.
78. Schumacher, E.F. 1973. *Small is beautiful: a study of economics as if people mattered*. London: Blond and Briggs Ltd.
79. Schwartz, Shalom H. 1996. "Value Priorities and Behavior: Applying a Theory of Integrated Value Systems." Chapter in The Psychology of Value. Clive Seligman, James M. Olsen, and Mark P. Zana eds, Lawrence Erlbaum Associates Inc., New Jersey.
80. Schwartz, Shalom H. "Universals in the context and structure of values: Theoretical advances and empirical tests in 20 countries." In: M.P. Zanna eds. Advances in experimental social psychology. Vol. 25, pp.1-65, Academic Press New York, 1992.
81. Seligman, Martin E.P., Mihaly Csikszentmihalyi. 2000."Positive Psychology. An Introduction." American Psychologist 55(1): 5–14.
82. Seligman, Martin E.P. 2002. *Authentic happiness: using the new positive psychology to realize your potential for lasting fulfillment*. New York: The Free Press.
83. Smith, Adam. 1776. *Wealth of nations* Digireads.com Publishers.
84. Snelders, Dirk H.M.J.J., and Stephen E.G. Lea. 1996. Different kinds of work, different kinds of pay: an examination of the overjustification effect. *Journal of Socio-Economics* 25(4): 517–535.
85. Snelders, Dirk H.M.J.J., Gonul Hussein, Stephen Lea, and Paul Webley. 1992. The polymorphous concept of money. *Journal of Economic Psychology* 13(1): 71–92.
86. Thøgersen, John, and Folke Ölander. 2002. Human values and the emergence of a sustainable consumption pattern: A panel study. *Journal of Economic Psychology* 23(5): 605–630.
87. Tischler, Len. 1999. "The growing interest in spirituality in business. A long-term socio-economic explanation". *Journal of Organizational Change* 12(4): 273–279.
88. Tukker, Arnold, Peter Eder, Martin Charter, Erick Haag, An Vercalsteren, and Thomas Wiedmann. 2001. Eco-design: The State of Implementation in Europe. Conclusions of a state of the art study for IPTS. *Journal of Sustainable Product Design* 1: 147–161.

89. UNDP. United Nations Development Reports: http://hdr.undp.org/en/reports/nhdr/. Accessed 3 January 2011.
90. De Valk, Maurice. 2010. Quality of the work life balance. The importance of values in well-being management. PhD dissertation. The Netherlands: Erasmus University Rotterdam.
91. Veenhoven, Ruut. 2010. Greater happiness for a greater number. is that possible and desirable? *Journal of Happiness Studies* 11: 605–629.
92. Veenhoven, Ruut. 1991. Is happiness relative? *Social Indicators Research* 24: 1–34.
93. Veenhoven, Ruut. 2000. The four qualities of life. Ordering concepts and measures of the good life. *Journal of Happiness Studies* 1(1): 1–39.
94. Wackernagel, Mathis, Larry Onisto, Patricia Bello, Alejandro Callejas Linares, Ina Susana Lopez Falfan, Jesus Mendez Garcia, Ma. Ana Isabel Suarez Guerrero, and Guadalupe Suarez Guerrero. 1999. National natural capital accounting with the ecological footprint concept. *Ecological Economics* 29: 375–390.
95. Wijnberg, Nachoem M. 2000. Normative stakeholder theory and aristotle: the link between ethics and politics. *Journal of Business Ethics* 25(4): 329–342.
96. Yunus, M., B. Moingeon, and L. Lehmann-Ortega. 2010. Building social business models: lessons from the grameen experience. *Longe Range Planning* 43: 308–325.

Chapter 4
Meaningful Innovation

Introduction

Tackling today's societal challenges requires a deeper level of innovation than just inventing products that sell well. Innovations will need to address the different levels of value to ensure benefits for society, the users and the different organizations involved in the ecosystem. To really address issues like the environment, increasing health care costs for an aging population or social coherence in crowded cities, new products and services will only have an impact if they also address a change of behavior. This is not likely to happen if people need to sacrifice value they have become used to, or which has been on their wish list for some time. Changing behavior is much more likely to happen if people recognize the need for change, because it is a pleasurable experience, or because it is fun to do so. It is certainly an option to design products so that people instinctively use them in a sustainable way [37].

The innovations we are aiming for will need to address user behavior, but are also likely to disrupt the current structures in the ecosystem. To give an example: if the shift from cars using fossil fuels to electric cars really happens, it will inherently reduce the power of the fossil fuel companies. This change will not happen overnight, but organizations may still find it difficult to follow the speed of adoption by users when attractive new products and services become available. To understand these innovations it is important to understand the links between the value generated at the different levels: users, organizations, ecosystems, and society. Figure 4.1 illustrates that such meaningful innovations combine the creation of value at all four levels. This chapter discusses the definition of new value propositions that create shared value for these four levels with the support of the Value Framework.

E. den Ouden, *Innovation Design*, DOI: 10.1007/978-1-4471-2268-5_4,

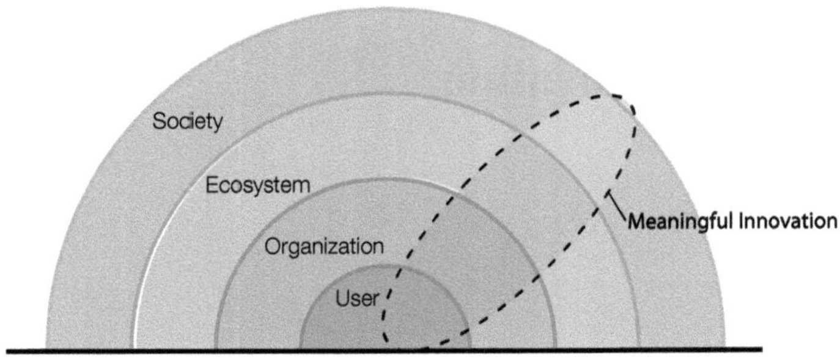

Fig. 4.1 Meaningful innovation: creating shared value for all four levels of the Value Framework

Defining New Value Propositions

A value proposition is defined in the business literature as a clear, simple statement of the tangible and intangible benefits of a new solution, together with an approximate price. It requires an understanding of a real customer with a fundamental problem in a given situation that needs a solution [34]. A value proposition should increase the benefits and/or reduce the sacrifices that customers perceive as relevant, build on the competences and resources that the company is able to utilize more effectively than its competitors, and be recognizably different (unique) from competition to create competitive advantage. The key dimensions of customer value to be addressed in the value proposition are: economic (price); functional (solutions); emotional (experience); and symbolic (meaning) [61].

Transformational innovations are different from regular products or services, in that they are often a system of related products and services that need to create value for a longer period: they are not like simple products that are purchased in a single transaction. Their extended use is crucial to achieve the sustained value that they are meant to deliver. Thus the value proposition for a transformational innovation should be crafted as an exchange of value that is described in terms of perceived benefits or reduced costs, shows clearly to whom that value should flow and how, is delivered over a time frame longer than a single transaction, is often co-created through interaction between two or more parties, and is broadly congruent with the relationship objectives set for a particular market [2].

Transformational innovations are also more risky than regular, incremental innovations. It is not known in advance whether the more radical solutions will be truly accepted by the users, and whether a viable business model and ecosystem can be created to sustain them. Thus in practice, the value proposition is created in an iterative process in which users and different stakeholders are co-creating together [14]. Emerging value for the different parties is revealed in a dialogue by using phrases like "if we…, will you…" to ensure that the value is reciprocal [3]. Value exchange in product and service systems is not just a direct exchange between a company and a customer, but between other stakeholders in the ecosystem as well [50]. The stakeholders in the ecosystem are all those groups and individuals who affect or are affected by the accomplishment of the innovation [24]. Thus the exchange of mutual beneficial value goes beyond the immediacy of the rather straightforward market transactions that are common in more traditional products or services sold directly to customers. In fact, in transformational innovations the opportunities for value creation are enhanced through the development of relationships that also takes place at the same time in the process. Value creation and relationship development are highly integrated. These developing relationships form the basis for creating and sustaining exchanges of value in the ecosystem [50].

Transformational innovations are in essence socio-technical service systems. Such systems include anything that is necessary to perform the intended function, including the environment and social context. The system consists of the products and services, but also explicitly includes the economic and social actors that are required in the collaboration to combine their individual offerings into a coherent, customer-facing solution [35]. The process of co-creating value propositions integrates three processes: the customer process—emotion, cognition, behavior; the supplier process—opportunities, planning, implementation; and the encounter process—communication, usage, service [51]. Value propositions for transformational innovation are therefore less straightforward than what is regularly described in the business literature.

A more elaborate description of value propositions for transformational innovations and how they can be described will be given in Chap. 5.

The Role of Innovation Of course, governments can define policies to achieve the necessary transformations. They can punish companies and citizens by taxing, fining, and prosecuting. And they can provide incentives to show the desired behavior through tax breaks and grants. There are many examples of these: tax levied on tobacco, increased tax benefits for cars that emit less carbon dioxide, fines for littering, grants for energy-saving solutions etc. For example, in the Europe 2030 project, opportunities for measures that governments can take are given to address the challenges [59]. Although this is important, not many readers of this book will be in a position to influence these processes and policies directly. What is more likely within the circle of influence is to create innovations that achieve the desired transformations. Disruptive innovations can undermine the current, ineffective, or even counterproductive structures, and make them obsolete or at least much less influential.

Another strategy could be to implement new technologies that contribute to meeting the societal challenges, without necessarily changing people's behavior. One example is changing from incandescent light bulbs to LED lamps. But quite often such technologies unintentionally change user behavior. In the case of energy-efficient light bulbs: 'people use them in places where they can leave them on for 24 h a day, because they use so little energy'. Similarly, an energy-efficient computer monitor will not deliver the intended energy savings if the user selects different settings, or leaves it switched on day and night. Companies can however, adopt a user-centered approach and guide users to behave in desired ways. Another example in the area of sustainability: Unilever has been working to increase the concentration of washing detergents with the aim of reducing the energy used in production and transportation. However, they realized that users tend to increase the amount of detergent used above what is needed for good washing results. Hence they decided to design a powder tablet specifically to counteract this 'rebound' effect. The tablet form prevents the use of excessive amounts of powder, and enables users to wash more sustainably [38].

Disruptive Innovations As already stated, the current systems may resist change because of the benefits they get from the current way of doing business. For example in health care, some organizations currently benefit from the existing structure that focuses much more on cure than on prevention. What is needed is an innovation that disrupts the current status quo. Disruptive innovations are based on very different value propositions than those that are currently available. They typically offer customers cheaper, simpler, smaller, and/or more convenient solutions [12]. Organizations or people that currently make a good living out of treating patients might resist the change to more preventive care if they do not see how they can still make a good living in the new situation. To be able to successfully implement disruptive innovations of this kind requires the creation of new ecosystems that bind together the organizations needed for overall success. In health care this includes insurance companies, health care service providers, medical professionals, manufacturers of medical products and systems for use in institutions and at home, and patient associations, to name just a few. Disruptive innovations will only succeed if sufficient stakeholders are willing to work together. In other words, sufficient value should be created to make these stakeholders adopt the idea and invest in it.

Example: the music industry

One example of an industry that has been shaken up by a disruptive innovation is the media. With the digitalization that has taken place in this industry, the positions formerly held by the traditional parties were suddenly no longer secure. When MP3 became a common music format, people could easily copy and share music through the Internet. This gave rise to new players that created online peer-to-peer sharing platforms, and new Internet shops from which people could legally download music. Initially there were a lot of illegal copies, as the early adopters of MP3 were computer-savvy and knew how to handle these digital platforms. For less computer-savvy users, platforms like Apple's iPod and iTunes were an easy, affordable, and legal way to follow this trend. The traditional business model of the music industry was based on relatively heavy investments in the printing of records and CDs that were distributed in mass markets. The costs involved meant that the record labels did a thorough screening of the potential of the artist. But digitalization means the costs involved are much lower, and MP3s can be created and made available on the Internet for downloading at very low cost. New platforms arose that allowed bands to upload music to platforms that allowed music enthusiasts to listen to it and indicate whether they liked it. The involvement of these music enthusiasts led to new 'hit parades' that gave unknown artists a chance. An example of such a platform is SellaBand.com. This internet-based platform manages all aspects of the value creation system, regardless of the artist's location, entrepreneurial orientation, and initial reputation. The system connects musicians from all over the world with music fans who can voluntarily turn into small investors in the production, distribution, and global success of their favorite music [19]. And today it is not uncommon for unknown talent to be discovered through completely open platforms like YouTube. All this means that the traditional role of the music industry has changed dramatically. New players have created completely new ecosystems, which to a large extent have displaced the traditional music industry. The revolution is not over yet. New ideas and new business models are still being generated. One of the newer ideas is to provide streaming music, which has become an option now that so many people are online 24/7. It is no longer necessary to have your own music collection if you can always access your favorite music online. Instead of spending money to buy a song, people can now subscribe to online music services. For a monthly fee they have access to huge music libraries. The Swedish company Spotify is one example of this new kind of service.

The examples show that roles and stakes can change dramatically if new innovations are able to create entirely new ecosystems. Players that were once dominant do not play a significant role in these new ecosystems. Newcomers can create a promising business if they understand what users want, and are able to translate this insight into value propositions that provide value for users.

The example of the music industry also shows that some of these new ecosystems are small, while others are significantly bigger. SellaBand is relatively small. The only thing that was needed to start SellaBand was a network of music lovers, a number of artists who were willing to upload music, and a web site. The investment needed to get started was relatively small, and since the service focused on new artists the established music corporations were not involved. Starting a service like iTunes is a different story. Because Apple wanted to make existing popular music available for legal downloading, they needed to make arrangements with the existing music corporations. Those big, global players regarded Apple as a serious candidate. Start-up companies might have had much more difficulty in being recognized and getting sufficient commitment from the corporations to invest time in a new proposition. But even for Apple the discussions on music rights management were an important issue. Not all music corporations were willing to share their music for downloading through iTunes—it took until December 2010 for iTunes users to be able to access the Beatles songs. The ecosystem for iTunes is also much larger than the one for SellaBand. iTunes had to create national shops, as regulations for digital rights and currencies differ between countries. For users it was important to be able to download their national artists, so even small music publishers have to be part of the ecosystem. And even the artwork has to be made available, as users also like to see the covers of the tracks and albums they have downloaded. Thus the whole process is much more complex than may appear at first glance. The examples show that both small start-ups and large, established companies can become part of total ecosystems, and that new value propositions do not necessarily require heavy up-front investments, but they do require larger ecosystems to thrive.

Transformational Innovations The transformational innovations that this book aims at are most likely to have larger ecosystems. If innovations aim to address the change of behavior of users to increase their quality of life, established systems are likely to be replaced by new ones. This means that all the relevant stakeholders need to be identified, and the value proposition has to provide value for these stakeholders as well as for users. This chapter discusses how value can be generated for the different stakeholders, building on the Value Framework as presented in Chap. 3.

The focus in this chapter is on how value can be created by transformational innovations. Each of the levels of value will be discussed. For users, the value of transformational innovations is translated into experiences that engage them in an enjoyable way. These innovations induce users to change their behavior, making them want to continue using the product or service and thereby contributing to an increased quality of life. For organizations that deliver (part of) the value proposition, the value of the innovations is translated into the opportunity to do well as an organization. The innovation should offer organizations a way to provide sustainable value to their customers, thereby ensuring the continuity of the organization. The ecosystem can only survive if the value proposition provides value to its members over a longer period of time. Adaptations of the value proposition will be needed in response to changes and dynamics taking place over time. The value of the proposition for the ecosystem is the ability to do good over time by enabling the creation of value for all the stakeholders, and thereby contributing to the quality of life of the users and the well-being of society. The value of the proposition for society is in its transformational power. If an innovation is successful it improves the quality of life for society as whole.

Value Propositions for the User

The core of transformational innovations is how users change their behavior to improve the quality of their lives. This effect can be either direct or indirect. An example of a direct improvement in the quality of life is an innovation that increases well-being immediately, such as lighting therapy for people suffering from winter depression. Users of this intense lighting during the darker periods of the year get increased levels of the cortisol hormone, and this directly influences

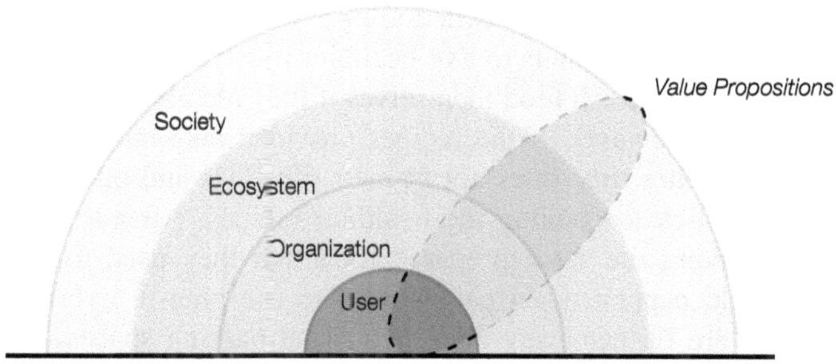

Fig. 4.2 Value propositions for the user

their energy level and mental well-being [36]. Indirect innovations include sustainable variants of existing products and services that provide the same functions but with less environmental impact. These innovations contribute indirectly and in the longer term to people's quality of life by maintaining a healthier living environment Transformational innovations should ultimately address both the direct, immediate influence and the indirect, longer term influence of the new solution (Fig. 4.2).

Transformational innovations need users to change their behavior, so that in a way these users become part of the product themselves. Pine and Gilmore call buyers of transformations 'aspirants'—they aspire to be someone or something different. Without a change in attitude, performance, characteristics, or some other fundamental dimension, no transformation can occur. Real transformations affect the very being of the user. If a change—losing weight, stopping a bad habit, reducing fixed expenses, or stopping wasteful practices—is only temporary and not sustained, then it was not really transformational but merely a momentary uptick along the same old journey. Likewise, any relapses or digressions reduce the measure or intensity of the transformation attained. With transformations, the customer is the product. When a company guides transformations, the offering is the individual. This means that the exact form and content of any particular transformational offering has to be considered carefully. The transformation elicitor must understand customer aspirations before hoping to affect any change in the particular traits—whether they are along physical, emotional, intellectual, or spiritual dimensions. Aspirations of course relate to customer expectations, but here the expectations center not on some external good or service, but on the customer himself: on what he wants to become [53].

Changing Behavior People are not likely to change their behavior if it requires a sacrifice. Some users might be willing to make a sacrifice to some extent if their motivational values are of the universalism kind. However, many more people are likely to change their behavior if a pleasurable experience is offered, and the innovation addresses their motivational values. It is essential, especially if a sustained change in behavior is needed, for the innovations to address people's values so that they will not be discarded after a while. As an example, if people need to change their eating habits to live healthier lifestyles, they might be willing to sacrifice by preparing fresh food themselves if they are given recipes that match their personal taste. However, if the recipes provided take more time to prepare and cook, and this extra time does not fit their lifestyles and other activities, not many people will stick to cooking the healthier recipes. After a while they will return to microwaving the same prepared food that they used to eat before. A similar effect could happen when people start to use plug-in hybrid cars. If the charging stations are further away than their usual parking spaces, and there are not enough of them, they may start waiting for a space to become vacant, or check back later in the evening to see whether it is available. But if too many attempts are unsuccessful after a couple of weeks, they might just decide to fill their tanks with gas again, and forget about electric power. So it is important that behavioral change is well supported in a way that fits in with people's everyday activities and with their values. What drives people and what is considered pleasurable is not the same for everybody, hence a thorough understanding of users' needs and values is needed.

Another example aimed to motivate teenage girls to be more physically active. The design that was created shows girls when their friends are walking, thereby leveraging the power of relationships for bonding-conscious teenage girls. They are less likely to reject friends' requests to go walking, so this product helps them to be more active and to meet their friends at the same time [70].

Target Behavior Many of the desirable societal transformations have an additional edge, in that people are not necessarily aware of the need to change their behavior. Persuasion has always been part of human experience, people have always sought to influence others' attitudes and behavior: through religious texts, training or telling stories [23], marketing [52], and more recently with computers [22]. Persuasive designs are created with the intention of changing behavior. Users of these designs are persuaded to change their attitudes and behavior through contextual information, advice, and guidance. The aim is to achieve a particular 'target behavior'. This can be influenced through interactions of a user with a system, or interactions between users and other users, mediated through a system. Lockton et al. [39] have created a list of target behavior strategies for both settings, with examples:

- The user follows a process or path, doing things in a sequence chosen by the designer, e.g. a customer places an order via a web site without missing out any steps;
- The user follows a process or path that is optimized for a particular situation, e.g. a user only spends as much time as is really needed in the shower;
- Decision among alternatives: a user's choice is guided, e.g. diners choose a healthier meal in an office canteen;
- Only certain users/groups of users can use something, e.g. only users who know their PIN can access their bank accounts at an ATM;

- Only users already behaving in a certain way get to use something, e.g. if a driver keeps to the speed limit the next set of traffic lights turn green, otherwise they stay red;
- No users can use something in a particular way, regardless of who they are or what they have done before, e.g a park bench fitted with a central armrest to prevent anyone from lying down;
- Users only get functionality when environmental criteria are satisfied, e.g. office lighting cannot be switched on if ambient daylight provides adequate lighting.

Different studies have looked into the role of goals and providing feedback information in achieving target behavior. It appears to be important that people can set their own goals, as opposed to being assigned them. People who choose their own goals are more committed to achieving them. Social and personality factors also influence the goals people set, and their commitment to achieving them. Pro-social individuals are more willing to accept costs to themselves in favor of societal gains than pro-self individuals. Nevertheless, a study of energy conservation in washing machines showed that self-set goals led to using 21% less energy, and assigned goals to 19.5%. Goal-setting, either by users themselves or assigned to them, was thereby shown to be a highly effective means of ensuring response to feedback. Just providing feedback without a goal being set was found not to be effective in encouraging behavioral change. The combination of a goal and effective feedback has proved to be a more effective way of encouraging people to change their behavior [42]. Persuasive technology needs to be perceived as subtle and supportive, and not as irritating, bullying, or patronizing. If the system is designed wrongly, people may become irritated as they get the same feedback not just once or twice but continuously, day in day out [32].

Behavior Model By designing experiences individuals and organizations aim to influence people's behavior. However, many attempts fail because of a lack of understanding of what factors lead to behavior change. Fogg [21] has developed a behavior model that identifies three factors that control whether behavior is adopted.

> For target behavior to happen, a person must have sufficient motivation, sufficient ability, and an effective trigger. All these three factors must be present at the same instant for the behavior to occur.

Motivation and ability are tradeoffs. People with a low motivation may perform a behavior if it is simple enough. On the other hand, people with a high motivation, might be creative in finding ways to increase their ability to achieve their goals. In many cases people are not at the extremes; they tend to have modest levels of motivation and ability. Effective persuasive technologies will boost either motivation or ability, e.g. by making the desired behavior easier, or both. Making a task simpler may include reducing the required time, making it cheaper, requiring less physical effort, requiring less thinking, making it less socially deviant, and embedding it in routine behavior.

The third factor in the model is the trigger. Successful triggers have three characteristics: they are noticed, they are associated with the target behavior, and they happen when the person is motivated and able to perform the behavior. This last element, timing, is crucial. If the trigger is presented at a moment when the person is not motivated (e.g. busy doing something else) or unable to respond (e.g. not having the means at hand), the trigger will not cause the target behavior to happen. Technology makes it easier to provide the trigger at the right moment. For example triggering to change the thermostat to lower temperature only when the person is at home.

Strategies Several strategies can be applied to seduce users into adopting specific behavior. As stated before, governments can introduce incentives and sanctions to guide citizens in their behavior. Although this may be effective it falls outside the scope of transformational innovations, as it is a strategy that is usually not within the influence of the designer or the organizations involved in the innovation.

Another option to change behavior is to try to make people feel guilty [37]. People indulging in unhealthy lifestyles like fast food or smoking are made to feel guilty and worry about their health. Education plays a role in making people aware of what the norm is, and making them feel guilty if they do not behave according to the accepted norms. Guilt, like shame, is a social motivator associated with social acceptance and rejection. In more collectivistic cultures, (avoiding) shame is an important driver. Being part of a group requires loyalty to the values of the group and being revealed as a person not behaving according to the social norm is a serious risk to people's personal reputation and therefore to their group as a whole. The effectiveness of guilt as a strategy to drive behavior has been debated, as past educational interventions have often failed. If people cannot relate emotionally or practically to the concept, they will not change their behavior. In many cases, the concepts that are presented to users are at a too high, macro-level—just like global warming, which is too far removed from people's daily experience. Also, the use of scientific language in the information that is provided does not appeal to most people. They lack the ability to 'feel' the impact of the practical decisions they need to make on these macro-level issues [38].

In the marketing discipline the issue of making consumers do things they do not want to do, in particular 'parting with their money', is a common phenomenon. Although there is a difference in making people do things they do not want to do once (as in a buying decision) with making them change their behavior sustainably, the insights are relevant in this context. Deighton and Grayson [16] distinguish seduction from persuasion. With persuasion there is preexisting consensus between the parties on the value of the transaction. In seduction, this preexisting consensus is lacking, and needs to be constructed in the interaction. The consumer must be moved, usually in stages, from old agreements to new. Seduction is different from fraud, where the customer becomes victim by being lured into an unsustainable and unfavorable deal. In seduction there is also a personal advantage for the consumer, but more effort is needed to induce him to enjoy things he did not intend to enjoy.

Design Designers can guide users in their behavioral change through the product design. There are several design strategies that can be applied.

To guide people towards sustainable behavior, three distinct strategies are defined in the literature [37, 73]:

- (Eco)-feedback: users are presented with specific information on the impact of their current behavior, and it is left to the users to relate this information to their own behavior and to change their behavior. An example in the case of littering is to label disposable coffee cups with information about how long they take to decompose.
- Scripting: the product is designed in such a way that the design triggers sustainable use, either by creating obstacles for unsustainable use, or by making sustainable behavior so easy that it is performed almost without thinking about it. In the case of littering, the understanding that users may not finish their soft drinks at once, and might want to retain the bottles for refilling with tap water, holds a strong case for a reclosable design. Such bottles are much less likely to end up as litter than tear-off closures that cannot be reused.
- Forced functionality: intelligent products and systems that adapt automatically to changing circumstances, or products with designed-in strong obstacles to prevent unsustainable behavior. An example of a forced functionality solution in the case of littering is the stay-on tabs on soft drink cans, that came into use in the 1980s as a replacement for loose tabs. By staying on the can, this tab design automatically prevents the sharp closure from being littered.

These three strategies differ in the amount of force that is used to guide the user. Feedback leaves the decision to change completely up to the user, scripts can be more or less forceful, and forced functionality leaves no option but to follow it.

A well-known example of forced functionality is office lighting systems with motion sensors. When everybody has left the room, the light is automatically switched off after 15 min or so. For such systems to be useful, it is important that they work well. The office lighting systems typically have difficulty in recognizing people who are working quietly at their desk. The sensors are often not sensitive enough to register small movements with the mouse or the flipping of a page when reading. If this happens, people may get annoyed and start to look for ways to 'hack' the system. This often leads to counter-effective solutions, which may even result in more energy use than a traditional, non-automated lighting system.

Another design strategy is to create engaging interactions. Overbeeke et al. [49] argue that this requires the restoration of the physicality of products, and the product design must address the user's action potential and capacity to appreciate sensory richness. Products must elicit rich interaction from the user. This opens up not only the functionality but also beauty and fun in interaction. Rich physical interaction offers even more opportunities; products might 'read' users' emotions and react to them in different ways.

There are several aspects involved in user engagement [62], but a common view is that engagement is intrinsically enjoyable. The activity is performed for its own intrinsic rewards, and not for extrinsic rewards [40]. The concept of engagement is referred to as a positive interactive state, in which people willingly give attention. During engagement people experience feelings of curiosity, interest, confidence and surprise [33]. A rich experience can be created by several means: external simulation gathered by the senses; internal stimulation gathered by the organs and muscles; and cortical stimulation via thought processes. There is also a relationship between the experience of richness and perceived control on the experienced engagement. An activity can require high or low skill levels, and can be perceived as a high or low challenge. Activities lead to the positive experience of 'flow', when high levels of both challenge and skills are experienced. If low levels of flow are experienced due to low levels of challenge and low levels of skill, interaction leads to feelings of apathy. In understanding an experience in interaction and creating engaging experiences, characteristics of the product, person, and the task should be taken into account simultaneously [62].

For effective change of behavior, design strategies of engaging and playful interactions are the best option as they seduce the user into a self-oriented, intrinsically rewarding activity [29]. People participate willingly in play because they enjoy it, or see the advantages of playing along [16].

The Need for Fact-Based, Up-to-Date Information To achieve the desired impact on society, it is crucial that transformational innovations guide users in the right direction. Therefore they need to be built on fact-based, up-to-date information. This information should be provided in an engaging way, which matches the level of interest of the specific users. For one user this may mean that he or she would like to get to know more about the background, hence additional information is desirable. For other users it means that they just want to know which option to choose, without being bothered with unnecessary information.

Providing the right information to seduce users can be a challenge in itself. In many cases there is no clear view (yet) on what the 'right' direction is. Especially when tradeoffs need to be made between incomparable dimensions the decision on what is right is ambiguous. Like the earlier example of wine: the tradeoff between the advantages of organically produced wine and additional transport distances.

Some 'context aware' innovations try to use information from the context of use to automatically adapt the system. This may also involve guiding the user through persuasion. Automatic generation of persuasive messages is not an easy challenge, as the physical world and human behavior are both complex and ambiguous [32]. Persuasive designs need to provide the relevant information for behavioral change in real time. If there is a delay in the information, people are less likely to remember it. Direct feedback helps in adjusting behavior. Redstrom gave an example of direct feedback on energy consumption. The aim of the 'erratic radio' is to show the user the overall energy consumption in the kitchen, and how energy use accumulates through the interactions with multiple objects. The erratic radio works like a normal radio, but also has a second layer of interaction. It detects electric fields emitted by active electronic appliances, and starts to detune and eventually lose its channel completely as energy use increases around it. To be able to use this radio, users have to pay attention to their energy consumption [58].

Ethical Considerations Even without the intention of persuading people to change their behavior, many innovations have had unintended or even accidental effects on people's behavior. The microwave oven is one example of this effect. Microwave ovens enable people to prepare meals quickly. The unintended side effect was that people started to change their eating behavior towards regularly and individually eating instant meals [72]. The traditional home-cooked meal, eaten together with the whole family, has largely been replaced by individual instant meals, which has surely had an impact on health and social contacts in families.

When designing transformational innovations, it is important to consider the intended as well as the unintended outcomes. With active and interactive seductive technologies increasing in power and impact, the moral and ethical dilemmas have increased significantly. Berdichevsky and Neuenschwander [4] created a flowchart that clarifies the level of ethical responsibility of designers. It basically means that a designer is not responsible for the outcome of an innovation when it was unintended and not reasonably predictable or when it was unintended but was reasonably predictable but the outcome is still ethical. The designer is considered praiseworthy when the outcome as ethical and intended. However there are also situations when the designer is considered at fault and responsible. The first situation is when the intended outcome is unethical. The second situation is when the outcome was not intended, but unethical and reasonably predictable.

Nowadays seductive technologies can build on an increasing amount of available information. For example information about users that is commonly available on social media like Facebook, or by gaining access to protected online information such as medical records. To illustrate that this is not without ethical and moral dilemmas, Berdichevsky and Neuenschwander use the example of monitoring water usage at a sink. If a child's water usage is monitored and the child is punished for leaving the tap running while brushing his teeth, the information is used for an immediate persuasive end. They termed this a 'little sister' technology. If a similar sink keeps track of whether restaurant employees wash their hands, so that their employer can later reward or punish them, and the same kind of information is collected and disseminated to a third party, they termed this a 'big brother' technology. If technologies are used to provide people with information about themselves, and relate to users one-to-one in a way that preserves the privacy of personal information, they are less likely to violate or infringe privacy. 'Little sister' technologies are therefore preferred over 'big brother' technologies.

Experience

Taking all the above into account, the challenge of creating new value propositions for transformational innovations at user level can be defined as:

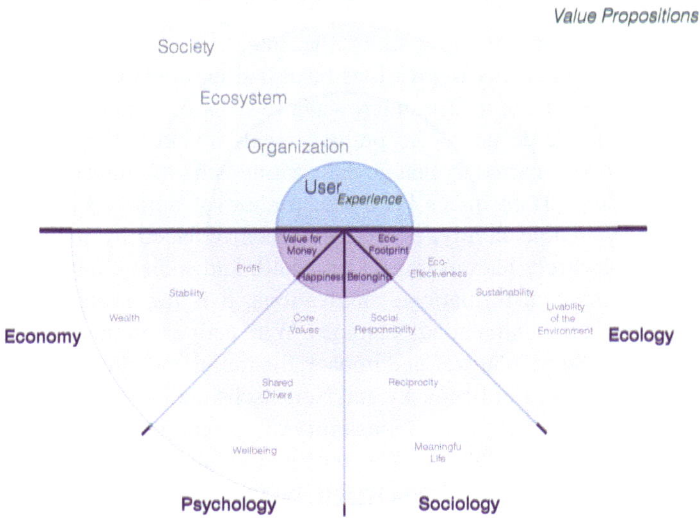

Fig. 4.3 Value propositions for the user: experience

> New value propositions will need to offer a pleasurable experience for the user, which seduces him or her to change behavior and keep on using the product or service to contribute to an increased quality of life.

Building on the Value Framework (see Fig. 4.3 for the specific elements at the user level) this challenge can be translated into more specific underlying goals.

The solutions presented to users need to offer value for money, so users can justify their purchase and extended use. They need to contribute to personal happiness by enabling users to pursue their individual dreams. They should enable users to belong to a group that is important to them. And last but not least, they should enable users to minimize their ecological footprint, so as to contribute to a livable planet.

An example of a product that provides users with a pleasurable experience and seduces them into changing their behavior for a longer period of time is the Nintendo Wii. The product offers games that help families to exercise together. Studies have shown that obesity is linked to lower activity levels associated with watching television and playing computer games, and the Wii aims to help solve this problem by making people more physically active [71]. It does not position itself as a competitor to the traditional computer games. In contrast to other computer games, it uses simple, intuitive 3D gestures to control the game, so people who are not familiar with traditional computer games, such as young children and the elderly, can use it straight away. It is sold at a price level below that of traditionally game consoles, enabling families with middle incomes to purchase it. The experience of doing physical activities together with friends or family in a game-like setting is experienced as fun and engaging. Other studies have shown that multi-player games increase the motivation to play interactive video games for an extended period [11]. Using the Wii has proved to be comparable to moderate-intensity physical activity, although not to the same extent as a real work-out [28, 37]. But there are also more negative reports, as some kids have found out that they can 'cheat' the system by making much smaller movements with the controller, and get symptoms of repetitive strain syndrome as a result. This unintended side effect shows the importance of understanding the users motivational values. It also shows that persuasion to make people do things they do not really like (such as physical activities for obese kids) has its limitations, as these users will apply all their creativity to find a 'work around'.

Value Propositions for the Organization

Transformational innovations offer organizations new business opportunities. Organizations that see these opportunities will actively start to work on innovations that provide solutions for societal challenges requiring transformation. These organizations take the lead in exploring the market opportunities, and in finding technologies that can be used in new solutions for products and services that bring value to the users.

It is important for an organization to find a superior value proposition: one that is clear to the users, and provides a simple statement of the tangible and intangible benefits; one that fits with the organization's strategy and enables the organization to provide real added value; one that provides the opportunity for a sustainable business and enables continuity of the organization.

As Hamel and Prahalad put it: winning in business is about getting to the future first [30]. This requires envisioning new markets, reinventing the organization, and following the path of the greatest opportunity. It is different from the traditional process of incremental product development that most organizations are more familiar with, and for which they have implemented more or less standard processes.

A number of issues related to creating new value propositions in an organization will be discussed in the following sections (Fig. 4.4).

End-User Driven Innovation Transformational innovations aim at developing business from creating true value for users. They address the unmet needs of users. Often they are based on recognizing needs that users did not even know they had, and they aim to provide solutions that users had not even imagined. These unmet needs provide the basis for new business opportunities. Uncovering unsolved problems, unvoiced needs, and market inefficiencies can lead to significant innovation opportunities. Organizations need to have the ability to see unmet needs not only before their competitors do, but often even before users are aware of them [65]. Traditional market research will not often be enough to identify such as yet unarticulated needs or emerging trends [68].

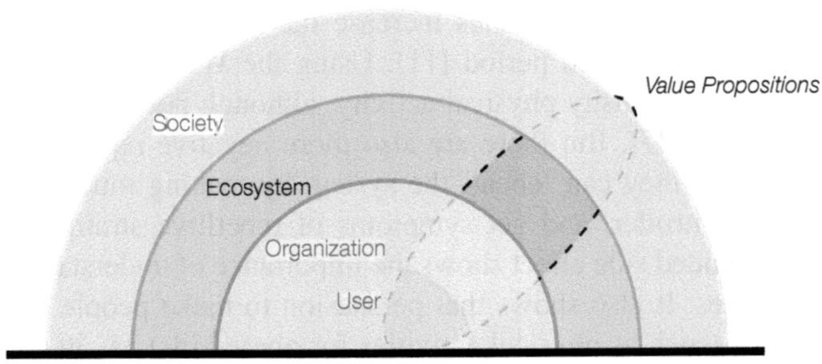

Fig. 4.4 Value propositions for the organization

End-user driven innovation is a very different approach from technology-driven innovation. While in the past technology was often the differentiating factor for companies to compete, today technologies are widely available from all kinds of suppliers, and the challenge lies much more in understanding the real unmet needs of users. If these needs are deeply understood, solutions can be built using any available technology. Sometimes a roadmap can be created that puts different technical options on a timeline, and links them to new market introductions of approved products. Technology is still important, but it is a means, and not an end in itself. Companies are also more flexible in choosing technologies because they can tap into what is available worldwide. Embarking on end-user driven innovation often leads to more collaborative product or service development with other organizations that bring in specific knowledge or technologies.

In the case of societal needs, the process of identifying unmet needs often means that organizations need to collaborate to bring in specific knowledge or experience, or to gain access to information or user groups. End-user-driven innovation then becomes a collaborative social process.

Social Innovation Social innovations are defined as innovative activities and services that are motivated by the goal of meeting a social need [44]. Social innovations are new concepts, adopted and utilized to resolve societal changes. Just as new products and processes are only considered to be 'innovations' when they become marketable, social innovations must bring a sustained benefit for their target audiences [8]. They therefore do not start from a specific company's business need or competitive strategy, but may be started by any organization that identified a social need. Social innovation (new ideas that improve quality of life and/or quantity of life) and business innovation (profitable new ideas) are different yet overlapping concepts. There are tremendous opportunities in these overlapping areas, in which business can go hand in hand with improving quality of life [55]; not in a philanthropic mode, because that would not be scalable, but by doing business with value propositions that improve quality of life.

In general these social innovations need the participation of a number of organizations to identify the unmet needs, and to generate the ideas for solutions that address them. This means that social innovations are often created in a more open setting with multiple organizations. The organizations that participate in such activities already have an open mindset. In many cases the starting point is a group of organizations that discuss a social need and then design a solution. They then realize that they need certain expertise to be added to the consortium, and start to approach other companies. These companies may not (yet) be so open-minded and in favor of open collaboration. The creation of such ecosystems around a value proposition will be discussed in more detail in Chap. 6.

Open Innovation Open innovation has become a widely adopted approach to ensure that ideas flow between different organizations and find their way to the market, even if the companies that invented them are not themselves involved in the market launch. Open innovation is mostly seen from the perspective of one organization [9, 10]. Open innovation can be seen as a step towards opening up the company's approach to innovation. Companies embarking on open innovation do so to allow ideas flowing into as well as out of the organization's innovation funnel. It looks into how an organization can benefit from ideas and technologies that are developed outside (inbound innovation) and how an organization can find external markets for ideas and technologies that are developed internally, but do not find their way to the companies' own markets (outbound innovation) [15].

If a company is market-driven, inbound open innovation is a logical approach as it widens the possible technological opportunities that can be used to create solutions. Applying open innovation itself is not sufficient for transformational innovations that should find their inspiration in societal needs. However, organizations that are closed, and struggle with the 'not invented here' syndrome, will have more difficulty to think outside their own means and open up to address social needs. Moreover, many organizations rely on analysis and prediction to take decisions in innovation, which is an ineffective approach in situations with a high degree of uncertainty. When dealing with a high degree of uncertainty, and inability to weigh information and ambiguous goals there cannot be a logic for decisions or choice. In such situations there is however, a logic of design, which starts with the means and then explores what goals can be achieved with these means with an entrepreneurial spirit [63].

Value Networks More and more research indicates that flexible value webs or value networks will replace the traditional static and linear value chains [1, 60, 66, 67]. These value networks typically involve different roles and organizations with different needs, hence value propositions are required to create value for the network of participating organizations.

Collaboration in networks has proved to be an important approach to create breakthrough innovations. Diverse collaborative networks have the capability of coming up with more novel solutions [45]. They are able to combine experience from different organizations and different viewpoints into one solution. Not only do more organizations together have more insights into the unmet needs of users, but they can also contribute different technologies and partial solutions to be integrated into a single complete solution.

The literature has proved the value of sharing knowledge and of collaborating in innovation, but that does not mean it is an easy step to take. Although most people have a desire to learn and to share what they know, in practice they behave differently [48].

Transformational innovations will be more selective in their choice of partners; not only do the partners need to bring in the required knowledge or assets to contribute to the innovation, but they will also need to have shared drivers and values. Value networks will be discussed in more detail in Chap. 6, which deals with the creation of ecosystems.

Self-Renewal Innovation itself is not new to many organizations. New product and service development processes are abundantly described in the literature, and many organizations have already understood the need to implement a structured innovation process to ensure sustainable business. However, transformational innovations are more radical than the traditional, more incremental innovations that most organizations are familiar with. To be able to make a success of these more radical innovations, organizations may need to renew themselves.

Self-renewal of an organization can be seen as a process of dissolving an existing organizational order and creating a new one. Order in an organization refers to the structural and cognitive order, which affects the pattern of the members of the organizational activities, namely, the pattern of resource deployment, organizational structure, system, processes, and cultures. As Nonaka indicates, there can be no self- renewal without dissolution and creation of order. People in the organization should have the freedom and autonomy to set their own goals. Through dialogue and discussion they create new points of view and integrate diverse individual perspectives into a new collective perspective [46].

Self-renewal means that conventional thinking needs to be challenged. Deeply held dogmas inside the organization and inside the industry about what drives success need to be challenged [65]. Orthodoxies about the right way to compete, price, organize and market, and the right business model, stifle progress and blind the organization to new opportunities. Or, as Govindarajan and Trimble put it [27]: 'If organizational memory is not tamed, it gets in the way of creation. Before you can create, you must forget.' The process of renewal will be much faster if the right people are on the team, and a team of great people will be much smarter in responding to changing conditions that are inevitably part of the process [13].

The potential of transformational innovations is to find new areas for growth. Even organizations that are currently experiencing growth will sooner or later run out of room to grow if they fail to reinvent themselves. Organizations find it hard to do this because they usually wait too long before realizing that the performance of their business is stalling. Organizations that have successfully reinvented themselves tend to have a broader scope. In essence they turn conventional wisdom on its head, and learn to focus on fixing what does not yet appear to be broken [47].

Virtuous Cycles Transformational innovations often require a new business model to create sustainable advantage, for example the opportunities in developing countries at the bottom of the pyramid require different business models to open up the market. Successful business models generate virtuous cycles that expand value creation and value capture over time. Examples of reinforcing cycles are found in Ryanair's business model. Although the reinforcing cycles of Ryanair are not linked to a social cause, and therefore cannot be considered true virtuous cycles, they provide a good example of how reinforcing cycles work. The business model rests on the key choices of offering customers low fares and providing nothing for free. Three reinforcing cycles are created [7]:

- Low fares = high volumes = greater bargaining power with suppliers = lower fixed costs = even lower fares;
- Low fares = high volumes = high aircraft utilization = low fixed cost per passenger = even lower fares;
- Low fares = expectations of low-quality service = no meals offered = low variable costs = even lower fares.

For transformational innovations, the reinforcing cycles need to take value beyond economic profit to be truly virtuous. Virtuous cycles are typically hard to start, hard to stop, and easy to maintain [5]. The reason that they are hard to start is that the platform that has to be created needs a critical mass of adopters, and a critical mass of complementary products and services. Once this has been achieved, the maintenance is easier. In the computer industry, as an example, it takes substantial new technological opportunities to get a new cycle going, because platform switches require large numbers of users to move from one platform to a new, incompatible, or only partially compatible one.

The virtuous cycles referred to above create sustainable businesses for the organizations involved. The virtuous aspect does not explicitly address the needs of users and society. For transformational innovations, it is important that the value proposition creates virtuous cycles that not only result in continuous value for the organization supplying the innovative products and services, but also for the users.

Organizations can create economic value by creating societal value. Porter and Kramer identify the virtuous circle of shared value: improving value in one area gives rise to opportunities in the others. Better connecting companies' successes with societal improvement will open up many ways to serve new needs, gain efficiency, create differentiation, and expand markets [56]. An example of such a virtuous circle is the monthly service for farmers in India, who pay a fee for weather and crop-pricing information and agricultural advice. The service reaches an estimated 2 million farmers, and research indicates it has helped increase the incomes of more than 60% of them—in some cases even tripling incomes. As innovations like these begin to work in poorer communities, new opportunities for economic development and social progress increase exponentially.

Building on Core Strengths Ideas for solutions in transformational innovations may be more or less close to the current business of an organization. Many organizations have analyzed what their core competences are. From the corporate perspective, these competences make a disproportionate contribution to customer-perceived value; they are competitively unique, and extendable; in a sense they provide a gateway to tomorrow's markets [30]. As competences take time to develop, most companies prefer a strategy that allows them to build on their core strengths to further develop themselves and find new business opportunities. Core competences are a strategic asset; they are difficult to imitate, develop, or acquire and that provides a basis for competitive advantage [65].

For transformation innovation the process starts the other way around: the core competences are not the starting point but the societal opportunity. It is important to understand the distance of this new opportunity from the core, as it is an indicator of risk and movement into the unknown. Although developing completely new competences to enter new markets is often perceived as too risky, an organization may need to diversify its offerings to create value with a new solution. Diversification can be more or less close to the core business. If it is further removed from the core the risks are higher, but the rewards may be higher too. Adjacency strategies are a way to plot a route to growth. They build on the strengths of the core business, and at the same time may serve to reinforce or defend the core. But they are also a journey into the unknown, a true extension of the core that pushes out the boundaries, and they are more risky than typical forms of organic growth [75].

When organizations consider embarking on more radical transformational innovations, it is important to understand the extent to which they expect to build on their core strengths. As transformational innovations involve relatively high risks, they typically follow an iterative process (see also Chap. 5), with insights that are continuously developing and a value proposition that is being adjusted accordingly. For organizations this means that the decision of whether an innovation is in line with their strategy, and of whether their role matches their core competences, is not one that can be made at the start of the project, but instead needs continuous monitoring. In an ideal situation, all organizations in the ecosystem will contribute to the overall value proposition from their core strengths.

Venturing Model It appears to be more difficult for larger organizations to embark on radical innovations. One reason for this is that for them to find new business attractive, the opportunity has to be either big or extremely profitable. But new markets are usually small, and by definition offer low margins. The paradox is that if organizations do not introduce new products, or enter new markets, the margins on their current products erode and they will eventually no longer be able to sustain the business. Thus for larger organizations innovations have to achieve two things: they need to provide a high enough margin to be of interest for the business manager, and they need to satisfy a large base of customers. A solution for this hurdle is to embark on corporate venturing [41]. Corporate venturing integrates different innovation approaches. It allows a portfolio of different activities, each with its own business dynamics. It may include incubation of new business activities and investments in existing but still maturing start-up enterprises. It is important though that organizations understand that these activities need to be managed differently from the existing, mature business. The decision making, rules, and processes need to be different. If the new business activities are treated in the same way as the existing, mature business, they will fail. New businesses need an atmosphere and attitude that is consistent with their specific needs. They will need to be able to take risk, use different approaches, explore new areas, and learn from failures. Experimentation is key to the success of a new business.

An issue that organizations struggle with is the extent to which they allow new business to cannibalize their existing business. The dilemma is that the existing business is often still profitable, and offers significant revenue. Most managers have the tendency to wait and further 'milk the cow' before embarking on a new business. This might be an option if there was no competition, but in today's competitive environment, if there is a serious opportunity for a new business it is highly unlikely that nobody else will notice it and take action to exploit it. In this situation, new and smaller entrants with no existing business to lose will often confront larger companies with disruptive innovations. To avoid this risk, venturing should be part of the research and development strategy of larger organizations. If larger organizations aim to play a role in ecosystems for new and transformational value propositions, it could be useful for them to approach their contributions internally as corporate ventures. The venturing process will be described in more detail in Chap. 5.

Doing Well

Thus, to summarize, the challenge in creating new value propositions for transformational innovations is to enable organizations to 'do well':

> New value propositions will need to provide an opportunity for sustainable value to ensure continuation of the organization.

This can be translated into more specific underlying goals with the help of the Value Framework. Figure 4.5 highlights the specific elements of the Value Framework at the organization level.

The new value proposition should provide a profitable business for the organization that markets (part of) the value proposition. The return on investment should be sufficient to cover the costs that are incurred. A good business model can contribute to profitability, but of course it requires the proposition to offer value to users and customers.

An organization that builds on its core competences will be better equipped to be successful with new products and services in a competitive arena. But the core values of an organization are important too: customers will better appreciate the product or service of a particular organization if it matches their perception of what the organization stands for. A value proposition should be authentic, in that what the organization offers matches how it is perceived by customers [26].

The new value propositions should explicitly address social responsibility and eco-effectiveness. These are not just additional requirements, but are the starting point of the innovation process. As stated earlier, the societal challenges are the

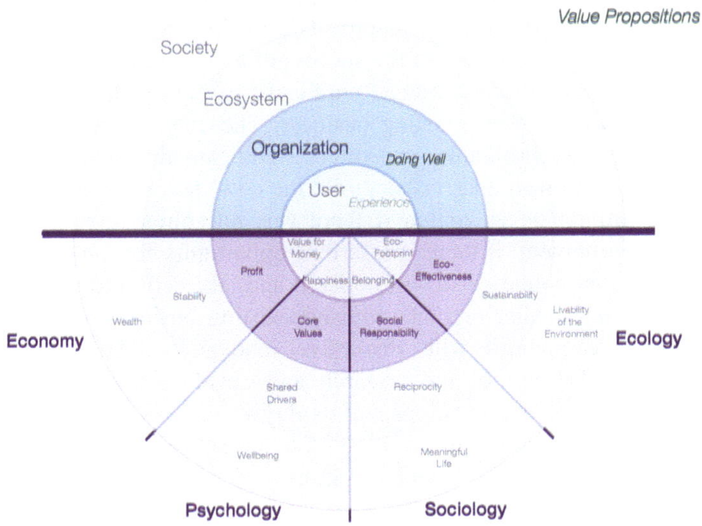

Fig. 4.5 Value propositions for the organization: Doing well

starting point of the innovation, not because organizations want to do a charitable favor to society, but because they provide tremendous business potential [56].

An example of an innovation that addresses all aspects at this level is the Philips SchoolVision lighting system. A system was created for classrooms, building on the opportunities of new lighting technologies to change the color and intensity of light. Depending on the type of class activities, the lighting can be adjusted to give bright, cool white light to help children concentrate early in the morning or shortly after lunch, or alternatively to give less intense, warmer white light to help them relax. The atmosphere in the classroom improves, and research shows that the children's learning capabilities also improve [54]. The new technologies also contribute to energy and cost reduction. For Philips the project supports the company's strategy and core value of improving people's lives through timely innovations.

Value Propositions for the Ecosystem

Transformational innovations typically require the creation of a new ecosystem that involves multiple stakeholders. Different stakeholders may be needed up-front in the creation of the innovation, because one company by itself does not have all the required knowledge and experience to deeply understand the unmet needs of the end-users and other stakeholders. Moreover, many innovations are a combination of various products and services from different companies, integrated into a

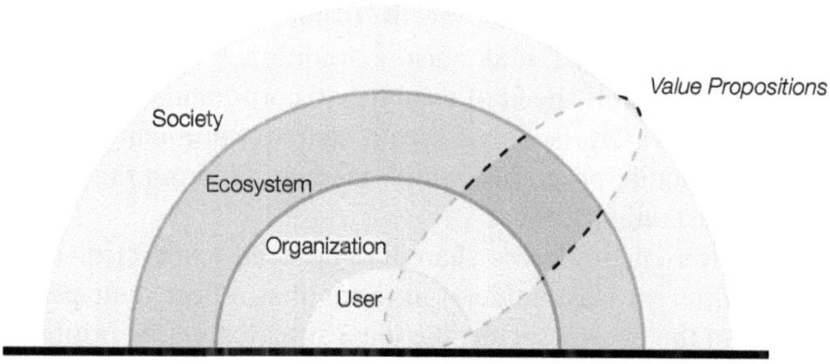

Fig. 4.6 Value propositions for the ecosystem

holistic solution that is launched on the market. The ecosystem then plays an important role in redistributing value among the different parties in the ecosystem. This is certainly the case when new business models are needed to ensure that the innovation is adopted. The new ecosystem is especially important if the innovation is disruptive; it needs to enable the establishment of an alternative to an existing, ingrained situation (Fig. 4.6).

Ecosystems are needed to ensure sustainable value for the different member organizations and individuals. The main objective of the ecosystem is to ensure that the value it creates for society, the users, and the organizations is sustainable. The ecosystem has an important role in 'doing good', and ensuring that this 'doing good' can be sustained over a longer period of time. The new value propositions for transformational innovations include the notion of the role of the ecosystem in creating robustness to ensure that the value generated keeps on flowing, despite unforeseen problems and changes that will need to be overcome over time.

Shared Value Porter and Kramer [56] recently introduced the term 'shared value'. Shared value creation focuses on identifying and expanding the connections between societal and economic progress. Shared value focuses on the right kind of profits—profits that create societal benefits rather than diminish them. The concept rests on the premise that both economic and social progress must be addressed using value principles. The essence of creating shared value with transformational innovations is to find new needs to meet, new products to offer, new customers to serve, and new ways to configure the value network. Competitive advantages that arise from creating shared value will often be more sustainable than conventional cost and quality improvements. The cycle of imitation by competition and reducing margins to zero to stay competitive can be broken.

From the perspective of society and the users, it does not matter what types of organizations create the value. What matters is that the benefits are delivered by those organizations—or combinations of organizations—that are best positioned to achieve the most impact for the least cost. The concept of shared value blurs the line between profit and non-profit organizations [56]. New kinds of hybrid enterprises are rapidly appearing. But it also blurs the line between large and small organizations. A whole generation of social entrepreneurs are pioneering new product concepts that meet social needs using viable business models [18]. Because they are not locked into narrow traditional business thinking, social entrepreneurs are often well ahead of established corporations in discovering these opportunities. Social enterprises that create shared value can scale up far more rapidly than purely social programs, which often suffer from the inability to grow and become self-sustaining [56].

Value propositions that address shared value need to be clear about what the value is for the different stakeholders, in terms that reflect their perspectives. To give an example in the energy sector: the ideal situation is that utilities make more money by pushing their users to save more electricity—hence the utility's total profits go up and the customer's total bills actually go down, because the energy savings more than offset the higher energy costs. An energy ecosystem that produces societal value (lower CO_2 emissions and energy efficiency) but not business value (great savings for customers and profits for utilities), is not scalable. It has to produce both. If the business model is created in a way that creates both values, it has the potential to disrupt the current situation, in which too many people have got rich for too long for doing the wrong things [25].

This aim of creating value for the different stakeholders is the core of the design of new ecosystems that will be discussed in Chap. 6.

Scalable Business Models Value propositions for transformational innovations should be based on unmet needs of users. When looking at society there is no shortage of unmet needs. In developing countries even the basic needs of many people are unmet [74]. But even in the developed world there are plenty of opportunities for innovation that improve nutrition, increase health, reduce energy use, or increase safety. These transformational value propositions in many cases also require a new business model [20], which has to ensure that the value proposition is delivered repeatedly and on the right scale. A sound business model makes the value proposition scalable. This is especially important for transformational innovations, because the positive impact of the innovation becomes larger as the number of users increases.

The business model should enable growth of the market. Many business models are built on economies of scale, as in the utilities, which favor larger producers over smaller ones [57]. Such business models usually need heavy investments, and they require a certain demand to become economically feasible. For the more uncertain, radical type of innovations, the risks of this model will often be too high. A much more iterative process of innovation is needed, starting with a small market to test the proposition and enrich it along the way so the market can grow. The business model has to enable this scaling; it needs to ensure a viable business in the early phases of market adoption, when quantities are still low, as well as in the mature phase, with large groups of intensive users of the products or services.

An example in a different context can be seen in the way Apple launched its App Store for the iPhone and iPod Touch. Most of the apps launched are initially free of charge. By enabling the use of the apps in larger numbers the developers can debug the apps and test their attractiveness to users much more quickly. In these early phases, they use customer feedback to improve the apps and add new features. Once they get the feeling that they have a compelling proposition, they start charging new customers for the app. If an app becomes a success it generates money for the developers and enables them to develop new and better apps. The business model enables the growth of the platform, which also benefits users as they are constantly provided with new and better apps. Apple gets part of the revenues of the apps sold, and can invest again in additional server capacity to run and improve the App Store platform itself, or in creating new features or new hardware like the iPad. In fact the App Store combines different business models: some apps are completely free of charge; others are free but generate different money flows by displaying advertisements; others are sold to users at a fixed price; while others are initially free with limited functionality, but offer additional functionality after a one-time investment; and again others are free, but charge for additional features that are regularly offered while using them.

A sound value proposition for transformational innovations is likely to consist of a combination of more than one business model, together providing scalability to the members of the ecosystem.

Robustness To provide lasting benefits to the species that depend on it, a biological ecosystem must persist in the face of environmental changes. Similarly, a business ecosystem should be capable of surviving disruptions such as an unforeseen technological change. The benefits are obvious: a company that is part of a robust ecosystem enjoys a relatively predictable environment, and the relationships among the members of the ecosystem are buffered against external shocks [31].

Innovation ecosystems, like biological ecosystems, are complex. There are multiple members, and the interdependencies are often numerous. A good understanding of the value proposition and how different members of the ecosystem contribute to the value of this proposition is crucial. Understanding the roles and interdependencies of the different members helps to judge the robustness of the system. Chapter 6 deals with new ecosystems and their robustness.

Doing Good

Thus the challenge in creating new value propositions for transformational innovations at ecosystem level is to enable 'doing good':

> New value propositions will need to allow the creation of ecosystems that can adapt to inherent changes and dynamics over a longer period of time and keep providing value for all stakeholders.

The value propositions need to be built on an ecosystem that enables 'doing good' in a sustainable way. Building on the Value Framework (see Fig. 4.7 for the

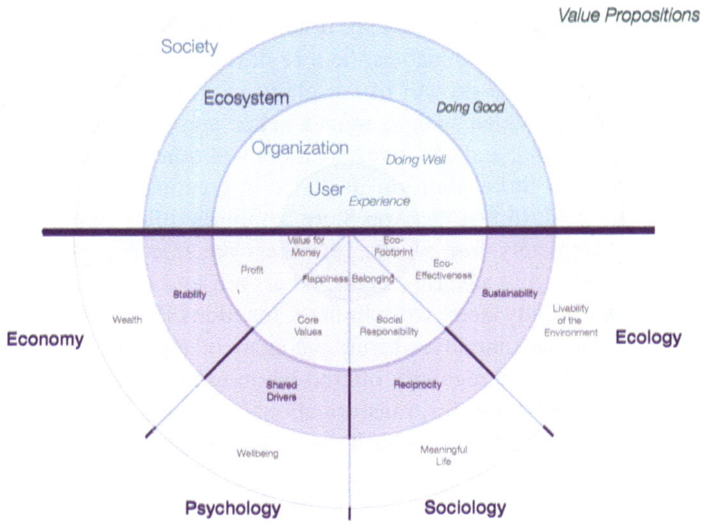

Fig. 4.7 Value propositions for the ecosystem: Doing good

specific elements at the ecosystem level) this can be translated into more specific underlying goals.

The value proposition should provide economic stability for the ecosystem, which means that the solution should deliver feasible economic value to the relevant stakeholders in the ecosystem. If the value proposition is not built on good economic value it will not survive long. As stated earlier, this may mean that the business model may have to change over time. But in the end, the value proposition will only be feasible if it can provide economic stability for the members of the ecosystem.

The value proposition should also be built on shared drivers of the important stakeholders. The members of the ecosystem have to agree on the overall purpose of the transformational innovation, and that purpose should drive their individual behavior. The motivation for the members of the ecosystem should always be to sustain that overarching goal, although progressively new insights, and changes in the environment may mean that the solution needs to be adapted to changing circumstances and advancing insights. Usually those changes do not impact the overall purpose of the innovation, but rather the solution that is provided or the way it is offered.

The value proposition should enable reciprocal relationships between the members of the ecosystem. This does not necessarily mean that all transactions in the ecosystem are directly reciprocal. Indirect reciprocity, e.g. through a third party, can be very common in an ecosystem. Also the 'currency' used in the transactions does not need to be the same—one member may invest time, another may invest knowledge, a third may invest money. But the important thing is that overall there is a balance in the system. All the important members should be able to have a good life in the ecosystem. Reciprocal relationships will support this principle.

Last but not least, the value proposition should enable the sustainability of the ecosystem in ecological terms. The advantage of embracing the notion of ecosystems for innovations is that society, the people and the planet, can be explicitly represented in the ecosystem. This will ensure that sustainable solutions are selected, and new opportunities to improve the sustainability of the solution over time can be adopted. As ecosystems are more capable of adapting to changing circumstances than single organizations, they are likely to perform better on sustainability. New value propositions can benefit from these possibilities.

Value Propositions for Society

Value propositions at societal level aim to enable social change: in general to improve the quality of life and overcome the challenges of the aging population, energy scarcity or any of the other challenges facing society. A transformation is needed because a lasting change needs to be anchored into the daily lives of the citizens (Fig. 4.8).

Social Innovation Social innovation was discussed earlier as a topic at the level of organizations: they can launch innovations that provide solutions to societal challenges. The purpose of social innovations is to fill a gap between what there is what there ought to be, between what people need and what they are offered by governments, private firms and non-governmental organizations. Mulgan identified some fields in which there are particularly severe innovation deficits, but also great opportunities for new creative solutions [43]:

- Climate change: which demands new thinking on how to reorder cities, transport systems, energy, and housing to dramatically reduce CO_2 emissions. Technology has a decisive role to play—but so will social innovations that help to change behavior.
- Growing diversity of countries and cities: which demands innovative ways of organizing schooling, language training, and housing to prevent segregation and conflict.
- Stark inequalities: which have widened in many societies, including the UK, US, and China, and tend to be associated with many other social problems, ranging from violence to mental illness.
- Rising incidence of long term conditions: such as arthritis, depression, diabetes, cancer, and heart disease (which are now chronic as well as acute), demanding novel social solutions as well as new models of medical support.
- Rising life expectancy: which requires new ways of organizing pensions, care and mutual support, new models of housing and urban design (for 4- and 5-generation families and continually changing housing needs), and new methods for countering isolation.

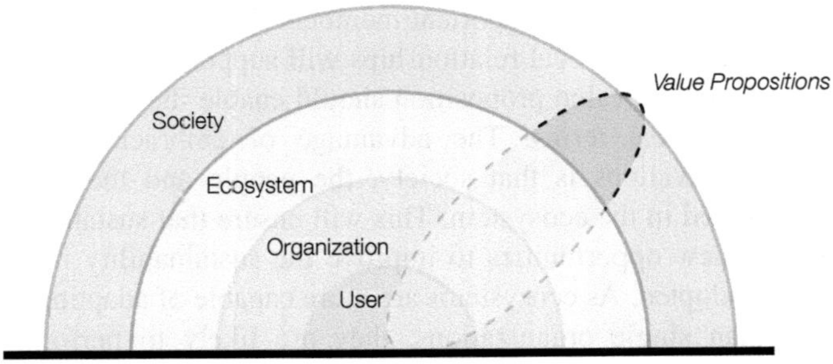

Fig. 4.8 Value propositions for society

- Behavioral problems and affluence: including obesity, bad diet and inactivity, as well as addictions to alcohol, drugs, and gambling.
- Difficult transitions to adulthood: requiring new ways for teenagers to successfully navigate their way into more stable careers, relationships, and lifestyles.
- Happiness: the mismatch between growing GDP, stagnant well-being, and declining real welfare according to some measures requires new ways of thinking about public policy and civic action.

In each of these fields the dominant existing models simply do not work well enough any longer. Often they are too inflexible and unimaginative. They may be fitted to past problems or bound by powerful interests. The result is unnecessary human suffering, and unrealized potential.

Realizing Lasting Change Although transformations require individuals to change their behavior, they only have an impact when large groups of people adopt the change sustainably. In the past the most far-reaching movements of change, such as feminism or environmentalism, involved millions of people and had dozens of intellectual and organizational leaders. Many of them also had the humility to realize that they were often as much following and channeling changes in public consciousness as they were directing them [43].

Change can be inspired in many ways, as discontent with the current situation can arise in many different places, from many different perspectives. What is important for change to have an impact is the process of spreading discontent among larger communities of people who are willing to take action. Nowadays social media can play an important role in quickly spreading information among large groups of individuals, who then form a community or movement for change.

An example of how movements can play a role is the Carrotmob campaign, named after using 'carrots' (rewards) instead of 'sticks' (punishment) to achieve change. On 29 March 2008, Carrotmob founder Brent Schulkin went to 23 convenience stores in San Francisco and persuaded them to make big environmental improvements in exchange for large numbers of new customers coming and spending money. Carrotmob was born when a 'mob' of hundreds of people came to the store with the biggest commitment, more than tripling its daily revenue in a few hours. Part of that revenue was then used to finance an energy-efficient retrofit of the lighting system. Carrotmob creates alignment between businesses, community organizers, and ordinary people in the community. Businesses are usually reluctant to 'do good' because doing so usually costs money, but they are happy to do good if they can make money in the process. Community organizers and activists frequently have trouble getting the attention of businesses they wish to change, but Carrotmob organizers have little trouble getting tangible results and mainstream support for their work. Many ordinary people do not get excited about participating in protests or boycotts, but they do like participating in fast, simple, mainstream, positive activity, which only requires them to spend money they would already spend anyway [6].

To create lasting change in user behavior, innovations are needed that have a lasting consequence beyond their consumption alone. As Pine and Gilmore already indicated: buyers of transformations seek to be guided towards some specific aim or purpose, and transformations must elicit that intended effect [53].

What is needed is that more and more people are inspired to embrace sustainable values and then to live those values through sustainable behavior. As Friedman puts it: laws and regulations tell you what you can do, but values tell you what you should do—there is a difference between doing what you have a right to do, and doing what is right to do [25]. Behavior is driven by people's basic values [69]. So to achieve for example environment-friendly behavior it is indeed important to ensure that people embrace sustainable values on a large enough scale.

Achieving lasting change in large groups of users is directly linked to the core of the Value Framework: creating an engaging experience for users. The challenge is to create this experience in such a way that it appeals to large groups of people, through a business model that is scalable.

Meaning As already indicated in Fig. 1.1, the mindset of people in the transformation economy is to be part of a community that addresses collective issues, and to make a meaningful contribution to that community. Worldwide, consumers are increasingly seeking products and services that connect them through meaning, which match their sense of how the world is, or should be. Diller et al. [17] created a list of meanings that people increasingly value (see Table 4.1).

There are numerous examples of people embarking on a project to contribute meaningfully to society, and who in very little time are able to attract other volunteers to get things going. An example of such a project initiated on an individual basis is the Plastic Whale project in the Netherlands. Marius Smit initiated the project because he was tired of world problems that are not really being solved either by international conferences that attract thousands of politicians, or by the promises of big organizations. To take up one problem, the problem of the 'plastic soup' (named after the large amounts of plastic waste in the Pacific Ocean and elsewhere), he reduces the problem to a practical level: how can you make a boat out of plastic waste? Despite having no background in design, plastic materials, sustainability or boats, he started by twittering and using

Table 4.1 List of meanings (alphabetical) [17]

Meaning	
Accomplishment	Achieving goals and making something of oneself; a sense of satisfaction that can result from productivity, focus, talent, or status
Beauty	The appreciation of qualities that give pleasure to the senses or spirit
Community	A sense of unity with others around us and a general connection with other human beings
Creation	The sense of having produced something new and original, and in so doing, to have a lasting contribution
Duty	The willing appreciation of oneself to a responsibility
Enlightenment	Clear understanding through logic or inspiration
Freedom	The sense of living without unwanted constraints
Harmony	The balanced and pleasing relationship of parts to a whole, whether in nature, society, or an individual
Justice	The assurance of equitable and unbiased treatment
Oneness	A sense of unity with everything around us
Redemption	Atonement or deliverance from past failure or decline
Security	Freedom from worry about loss
Truth	A commitment to honesty and integrity
Validation	The recognition of oneself as a valued individual worthy of respect
Wonder	Awe in the presence of a creation beyond one's understanding

LinkedIn contacts to tell people about his dream of sailing a boat made of plastic waste through the Netherlands in June 2011. In just a couple of weeks he was able to collect a whole crowd of people who are contributing knowledge, ideas, suggestions, time, or sponsoring to the project.

Shared Value Shared value has already been discussed in the section on value propositions for ecosystems. The reason to discuss it here again is that the concept of shared value is important to link value for society to value for organizations, and last but not least to value for users (who are also citizens of society). Another term used in the literature is linked-benefit strategies. Linked-benefit strategies combine the design strategies to create sustainable and commercially attractive products with design strategies for sustainable behavior, thereby minimizing user-related environmental losses. They link sustainable benefits to financial, functionality, aesthetic, and/or convenience benefits [64]. Transformational innovations will only become a real success and will only have a significant impact on society if the benefits to all parties involved are clear.

Transformation

Thus the challenge in creating new value propositions for transformational innovations at societal level is to achieve a transformation:

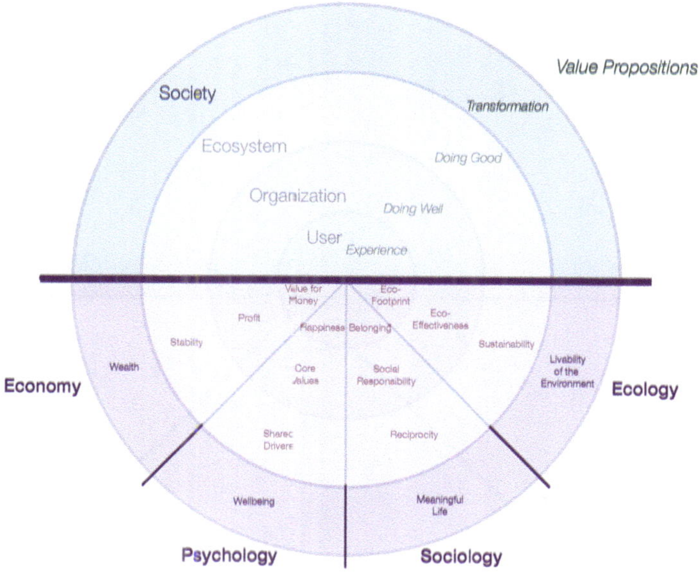

Fig. 4.9 Value propositions for society: Transformation

New value propositions will need to improve the quality of life for society as a whole and care for people and planet.

The value propositions need to promote societal transformation. Building on the Value Framework (see Fig. 4.9 for the specific elements at the societal level), this can be translated into more specific underlying goals.

The new value proposition should create economic wealth for society. Of course, this aspect is less relevant for developed countries than for developing countries, but society at large can still benefit from wealth generated by transformational innovations. Wealth creates opportunities to invest in aspects like the well-being of people and planet. Aiming to create wealth through new value propositions also distinguishes them from purely philanthropic activities that are usually less successfully scalable and gain limited support in the long term, as the money required cannot be used for other purposes. Creating wealth is instrumental in making sure that an innovation can flourish in the long term.

New value propositions for societal transformation should explicitly address the well-being of society, which is a community of individual people, as well as enabling people to contribute to society and have meaningful lives. There is a direct link between the aspirations of individual users (the inner circle of the Value Framework) and society, which is in essence a large social community of individual people. When transformational innovations are really successful they create value for large numbers of citizens, either directly because they are users of the innovation, or indirectly because they are stakeholders in the larger ecosystem, and benefit from the wealth and welfare that the innovation brings to society at large.

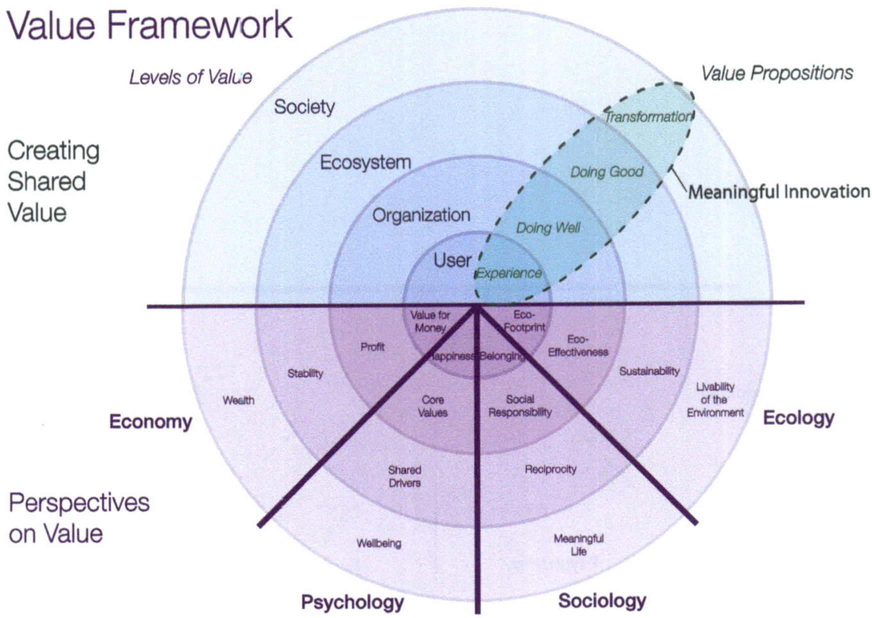

Fig. 4.10 Value propositions for shared value for all important stakeholders: meaningful innovations

Last but not least of the aspects in the model is the livability of the environment. Transformational innovations can contribute to ensuring that we care for people and planet, and find ways to accommodate the growing population without damaging the environment (and preferably even restoring the damage done in the past).

Meaningful Innovation

New value propositions for transformational innovations can be initiated at any level, and from any perspective—whether it is a user complaint about the value for money of a certain product, a group in society that aims at a more meaningful life, or a company that wants to improve its social responsibility. Each of the segments in the Value Framework can be the inspiration point for a new value proposition.

New value propositions can address one or more segments of the Value Framework. The more segments they cover, the more value they create. Ultimately, an innovation covers all 16 segments. By doing so it combines a positive user experience with 'doing well' and 'doing good', and enables an important transformation in society. Such an innovation is truly meaningful if it is built on shared value for all the important stakeholders at all levels. This is shown by the ellipse in Fig. 4.10: meaningful innovation crosses all levels of value, and creates shared value for all these levels at the same time. The process to develop meaningful innovation will be further discussed in Chap. 5.

The Value Framework helps to identify additional value opportunities to enrich an idea for a new value proposition, but it can also provide inspiration to create meaningful innovations.

References

1. Allee, Verna. 2008. Value network analysis and value conversion of tangible and intangible assets. *Journal of Intellectual Capital* 9(1): 5–24.
2. Ballantyne, David, and Richard J. Varey. 2006. Creating value-in-use through marketing interaction: The exchange logic of relating, communicating and knowing. *Marketing Theory* 6(3): 335–348.
3. Ballantyne, David, Pennie Frow, Richard J. Varey, and Adrian Payne. 2010. Value propositions as communication practice: Taking a wider view. *Industrial Marketing Management* 40(2): 202–210.
4. Berdichevsky, Daniel, and Erik Neuenschwander. 1999. Towards an ethics of persuasive technology. *Communications of the ACM* 42(5): 51–58.
5. Bresnahan, Timothy F. 1999. New modes of competition: Implications for the future structure of the computer industry. In *Competition, convergence, and the microsoft monopoly: Antitrust in the digital market place*, eds. J.A. Eisenach, and T.M. Lenard. Boston, MA: Kluwer Publishers.
6. Caplan, Jeremy. "Shoppers, Unite! Carrotmobs are Cooler Than Boycotts". Time Magazine, 15 May 2009.
7. Casadesus-Masanell, Ramon, and Joan E. Ricart. 2011. How to design a winning business model. *Harvard Business Review* 89(1–2): 100–107. January.
8. Centre for Social Innovation. "Stimulating Social Development." Published by the Centre for Social Innovation, ZSI Discussion Paper, no. 10. Vienna, August 2008.
9. Chesborough, Henry. 2003. *Open innovation. The new imperative for creating and profiting from technology*. Boston, MA Harvard Business Press.
10. Chesborough, Henry, Wim Vanhaverbeke, and West Joel (eds.). 2006. *Open innovation: Researching a new paradigm*. New York: Oxford University Press.
11. Chin, A. Paw M.J., W.M. Jacobs, E.P. Vaessen, S. Titze, and W. Van Mechelen. 2008. The motivation of children to play an active video game. *Journal of Science and Medicine in Sport* 11(2): 163–166.
12. Christensen, Clayton M. 1997. *The innovators dilemma: When new technologies cause great firms to fail*. Boston, MA: Harvard Business Press.
13. Collins, James C. 2001. *Good to great: Why some companies make the leap–and others don't*. New York: Harper Business.
14. Cova, Bernard, and Robert Salle. 2008. Marketing solutions in accordance with the S-D logic: Co-creating value with customer network actors. *Industrial Marketing Management* 37(3): 270–277.
15. Dahlander, Linus, and David M. Gann. 2010. How open is innovation? *Research Policy* 39(6): 699–709. July.
16. Deighton, John, and Kent Grayson. 1995. Marketing and seduction: Building exchange relationships by managing social consensus. *Journal of Consumer Research* 21(4): 660–676.
17. Diller, Steve, Nathan Shedroff, and Darrel Rhea. 2006. *Making meaning: How successful businesses deliver meaningful customer experiences*. Berkeley: New Riders.
18. Ellis, Tania. 2010. *The new pioneers: Sustainable business success through social innovation and social entrepreneurship*. London: Wiley.

19. Etemad, Hamid, Ian Wilkinson, and Leo Paul Dana. 2010. Internetization as the necessary condition for internationalizing in the newly emerging economy. *Journal of International Entrepreneurship* 8: 319–342.
20. Eyring, Matthew J., Mark W. Johnson, and Hari Nair. 2011. New Business models in emerging markets. *Harvard Business Review* 89(1/2): 88–95. January.
21. Fogg, B.J. "A Behavior Model for Persuasive Design." Persuasive '09, Claremont, California, USA, April 26–29, 2009.
22. Fogg, B.J. 2002. *Persuasive technology. Using computers to change what we think and do.* Waltham, MA: Morgan Kaufmann Publishers.
23. Fogg, B.J. 1999. Persuasive technologies. *communications of the ACM* 42(5): 26–29.
24. Freeman, R. Edward. 1984. *Strategic management: A stakeholder approach.* Boston: Pitman.
25. Friedman, Thomas L. 2009. *Hot, flat and crowded: Why the world needs a green revolution and how we can renew our global future.* London: Penguin Books.
26. Gilmore, James H., and B. Joseph Pine II. 2007. *Authenticity: What consumers really want.* Boston, MA: Harvard Business Press.
27. Govindarajan, Vijay, and Chris Trimble. 2011. The CEO's role in business model reinvention. *Harvard Business Review* 89: 108–114. Jan/Feb.
28. Graves, Lee E.F., Nicola D. Ridgers, and Gareth Stratton. 2008. The contribution of upper limb and total body movement to adolescents' energy expenditure whilst playing nintendo wii. *European Journal of Applied Physiology* 104: 617–623.
29. Grayson, Kent. 1999. The dangers and opportunities of playful consumption. In *Consumer value*, ed. Morris Holbrook, 105–125. New York: Routledge.
30. Hamel, Gary, and C.K. Prahalad. 1994. *Competing for the future: Breakthrough strategies for seizing control of your industry and creating the markets of tomorrow.* Boston, MA: Harvard Business Press.
31. Iansiti, Marco, and Roy Levien. 2004. *The keystone advantage: What the new dynamics of business ecosystems mean for strategy, innovation, and sustainability.* Boston, MA: Harvard Business Press.
32. IJsselstein, Wijnand, Yvonne de Kort, Cees Midden, Berry Eggen and Elise van den Hoven. 2006. Persuasive technology for human well-being: Setting the scene. In *Persuasive 2006*, Lecture Notes in Computer Science, 3962, 1–5. Berlin: Springer-Verlag.
33. Jacques, R., J. Preece, and T. Carey. 1995. Engagement as a design concept for multimedia. *Canadian Journal of Educational Communications* 24(1): 49–59. January.
34. Johnson, Mark W., Clayton M. Christensen, and Henning Kagermann. 2008. Reinventing your business model. *Harvard Business Review* 86(12): 50–59. December.
35. Kroes, Peter, Maarten Franssen, Ibo van de Poel, and Maarten Ottens. 2006. Treating socio-technical systems as engineering systems: Some conceptual problems. *Systems Research and Behavioral Science* 23(6): 803–814.
36. Kueller, Rikard. 2002. The influence of light on circarhythms in humans. *Journal of Physiological Anthropology and Applied Human Science* 21(2): 87–91.
37. Lanningham-Foster, Lorraine, Randal C. Foster, Shelly K. McCrady, Teresa B. Jensen, Naim Mitre, and James Levine. 2009. Activity promoting games and increased energy expenditure. *Journal of Pediatrics* 154(6): 819–823.
38. Lilley, Debra, Vicky Lofthouse, and Tracy Bhamra. 2005. Towards instinctive sustainable product use. 2nd International Conference: Sustainability Creating the Culture, Aberdeen Exhibition and Conference Centre, Aberdeen, 2–4 November.
39. Lockton, Dan, David Harrison, and Neville A. Stanton. 2010. The design with intent method: A design tool for influencing user behaviour. *Applied Ergonomics* 41(3): 382–392.
40. Malone, T.W. 1981. Toward a theory of intrinsically motivating instruction. *Cognitive Science* 4: 333–369.
41. Mason, Heidi, and Tim Rohner. 2002. *The venture imperative: A new model for corporate innovation.* Boston, MA: Harvard Business Press.

42. McCalley, L.T., and C.H.J. Midden. 2002. Energy conservation through integrated product-feedback: The roles of goal-setting and social orientation. *Journal of Economic Psychology* 23(5): 55–70.
43. Mulgan, Geoff. "Social Innovation. What it is, why it matters, and how it can be accelerated." Working paper from the Skoll Centre for Social Entrepreneurship. 2007
44. Mulgan, Geoff. 2006. The process of social innovation. *Innovations* 1(2): 145–162. Spring.
45. Nieto, Maria Jesus, and Lluis Santamaria. 2007. The importance of diverse collaborative networks for the novelty of product innovation. *Technovation* 27: 367–377.
46. Nonaka, Ikujiro. 1991. The knowledge-creating company. *Harvard Business Review* 69: 96–104. November.
47. Nunes, Paul, and Tim Breene. 2011. Reinvent your business before it's too late. *Harvard Business Review* 89(1/2): 80–87.
48. O'Dell, Carla, and C. Jackson Grayson. 1998. If only we knew what we know: Identification and transfer of internal best practices. *California Management Review* 40(3): 154–174.
49. Overbeeke, Kees, Tom Djajadiningrat, Caroline Hummels, Stephan Wensveen, and Joep Frens. 2003. In *Funology: From usability to enjoyment*, eds. Mark A. Blythe, Andrew F. Monk, Kees Overbeeke, and Peter C. Wright. The Netherlands: Kluwer Academic Publishers.
50. Payne, Adrian, David Ballantyne, and Martin Christopher. 2005. A stakeholder approach to relationship marketing strategy: The development and use of the "six markets" model. *European Journal of Marketing* 39(7/8): 855–871.
51. Payne, Adrian F., Kaj Storbacka, and Pennie Frow. 2008. Managing the co-creation of value. *Journal of the Academy of Marketing Science* 36: 83–96.
52. Petty, R.E., and J.T. Cacioppo. 1981 *Attitudes and persuasion: Classic and contemporary approaches*. Dubuque, IA: Wm. C. Brown.
53. Pine, B.Joseph, and James H. Gilmore. 1999. *The experience economy work is theatre and every business a stage*. Boston, MA: Harvard Business Press.
54. Philips SchoolVision. Press release (in Dutch): http://www.newscenter.philips.com/pwc_nc/main/shared/assets/nl/Newscenter/2010/SchoolVision/persbericht_schoolvision_final.pdf.
55. Pol, Eduardo, and Simon Ville. 2009. Social innovation: Buzz word or enduring term? *Journal of Socio-Economics* 38: 878–885.
56. Porter, Michael E., and Mark R. Kramer. 2011. Creating shared value. *Harvard Business Review* 89(1/2): 62–77. January.
57. Rappa, M.A. 2004. The utility business model and the future of computing services. *IBM Systems Journal* 43(1): 32–42.
58. Redstrom, Johan. 2006. Persuasive design: Fringes and foundations. In *Persuasive 2006*, ed. IJsselsteijn, Wijnand, Yvonne de Kort, Cees Midden, Berry Eggen, and Elise van den Hoven, Lecture Notes in Computer Science, 3962.
59. Reflection Group on the Future of the EU 2030. "Project Europe 2030. Challenges and Opportunities." A Report to the European Council, May 2010.
60. Riedl, Christoph, Tilo Boehmann, Jan Marco Leimeister and Helmut Krcmar. "A framework for analysising service ecosystem capabilities to innovate." 17th European Conference on Information Systems ECIS, Verona, Italy, 2009.
61. Rintamaki, Timo, Hannu Kuusela, and Lasse Mitronen. 2007. Identifying competitive customer value propositions in retailing. *Managing Service Quality* 17(6): 621–634.
62. Rozendaal, M.C. Designing engaging interactions with digital products, PhD thesis, Faculty of Industrial Design Engineering, Delft University of Technology, 2007.
63. Sarasvathy, Saras D. 2008. *Effectuation. elements of entrepreneurial expertise*. Northampton, MA: Edward Elgar Publishing Ltd.
64. Schmalz, Johannes, and Casper Boks. "Sustainable, user behaviour centered design applying linked-benefit strategies." The Logi Desk Lamp. 14th European Rondtable on Sustainable Production and Consumption, Delft, The Netherlands, October 25–29, 2010.
65. Skarzynski, Peter, and Rowan Gibson. 2008. *Innovation to the core: A blueprint for transformational the way your company innovates*. Boston, MA: Harvard Business Press.

66. Stathel, Stephan, Jan Finzen, Christoph Riedl, and Norman May. "Service innovation in business value networks". XVIII International RESER Conference, Stuttgart, 2008.
67. Tapscott, D., A. Lowy, and D. Ticoll. 2000. *Digital capital: Harnessing the power of business webs.* Boston, MA: Harvard Business Press.
68. Teece, David J. 2010. Business models, business strategy and innovation. *Long Range Planning* 43: 172–194.
69. Thøgersen, John, and Folke Ölander. 2002. Human values and the emergence of a sustainable consumption pattern: A panel study. *Journal of Economic Psychology* 23(5): 605–630.
70. Toscos, Tammy, Anne Faber, Shunying An, and Mona Praful Gandhi. 2006. Chick clique: Persuasive technology to motivate teenage girls to exercise. Proceedings of CHI 2006, Montreal, Quebec, Canada, April 22–27.
71. Vandewater, E.A., M.S. Shim, and A.G. Caplovitz. 2004. Linking obesity and activity level with children's television and video game use. *Journal of Adolescence* 27(1): 71–85.
72. Verbeek, Peter-Paul. 2006. Acting artefacts: The technological mediation of action. In *User behaviour and technology development: Shaping sustainable relations between consumers and technologies*, eds. P.-P. Verbeek, and A. Slob, 53–60. Berlin: Springer.
73. Wever, Renee, Jasper van Kuijk, and Casper Boks. 2008. User centered design for sustainable behaviour. *International Journal of Sustainable Engineering* 1(1): 9–20.
74. Yunus, M., B. Moingeon, and L. Lehmann-Ortega. 2010. Building social business models: Lessons from the grameen experience. *Longe Range Planning* 43: 308–325.
75. Zook, Chris. 2004. *Beyond the core: Expand your market without abandoning your roots.* Boston, MA: Harvard Business Press.

Chapter 5
Creating Meaningful Value Propositions

Introduction

The first part of this book had a more theoretical and reflective character. This second part makes a transition to more practical information: how to get going with the definition and realization of meaningful innovations. This chapter makes a start on creating new value propositions for meaningful innovations. Meaningful innovations, as introduced in the previous chapter, combine value at the four levels of the Value Framework, see also Table 5.1.

Meaningful innovations aim to create a more holistic value, in which the value of the whole is perceived as more than the sum of its parts. This requires integrative thinking, as the seemingly conflicting needs of different stakeholders will need to be integrated into one overall value proposition.

Starting Points for New Value Propositions Starting points for new value propositions can be found in any of the sixteen segments in the Value Framework. Figure 5.1 shows some examples of starting points for new value propositions. The examples are just to illustrate that each of the segments can be an inspiration for innovations, the examples are not intended to be exhaustive, and neither are they necessarily the best examples of innovations that address all levels of value.

For companies it is very logical to start projects to generate profit with the new products and/or services. Such projects can aim for more or less societal relevance. For example, Philips created the Ambilight TV as a result of the idea that people would like to unwind after an intensive day at work. One way that a lot of people like to do this is by watching a movie. Light effects on the wall were added to a TV screen to increase the perception of a full immersive experience. This additional value for users allowed the company to sell the product in a higher price segment. For social entrepreneurs, the deprivation of poor people in society can be a starting point to look for ways to improve their wealth and well-being, as is the case with Grameen Bank in Bangladesh [85]. A design agency may start from an unmet user need and the ambition to create a pleasurable user experience, like the famous

E. den Ouden, *Innovation Design*, DOI: 10.1007/978-1-4471-2268-5_5,

Table 5.1 Elements of meaningful innovations

Level	Value proposition	
User	Experience	Offers a pleasurable experience for users, seducing them into changing their behavior and keeping on using the product or service to contribute to an increased quality of life
Organization	Doing well	Provides an opportunity for sustainable value to ensure the continuity of the organization
Ecosystem	Doing good	Allows the creation of ecosystems that can adapt to inherent changes and dynamics over a longer period of time and keep providing value for all stakeholders
Society	Transformation	Improves the quality of life for society as a whole and cares for people and planet

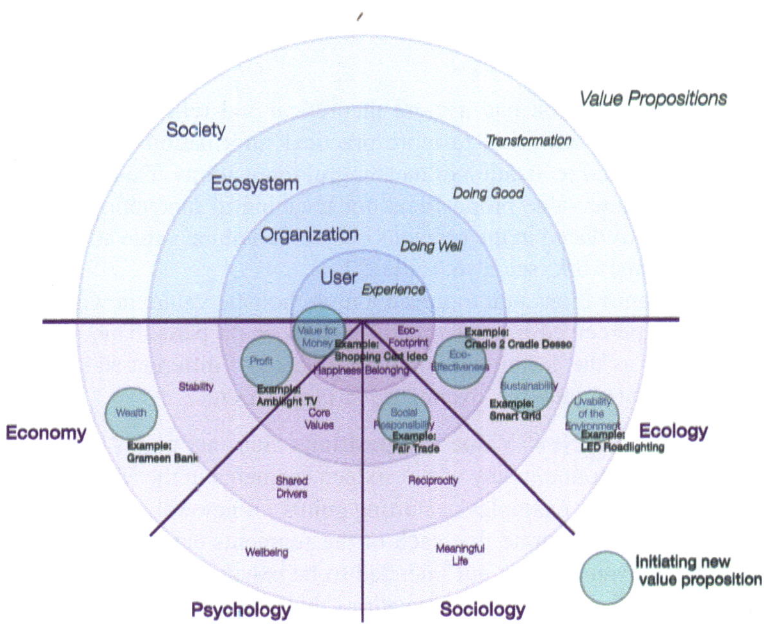

Fig. 5.1 Starting points for new value propositions—some examples

IDEO project to redesign a supermarket shopping cart [41]. A company may adopt a Fair Trade policy to improve the living conditions of coffee farmers in developing countries [55]. Or a company may decide to implement Cradle-to-Cradle to decrease the ecological footprint of its business, such as Desso which decided to redesign its carpets and manufacturing processes toward completely renewable and recycled sources [48]. A municipality may want to increase the livability of the environment and switch to dynamic LED lighting for paths in a park where plants and animals would otherwise be disturbed by the continuous bright light. Or

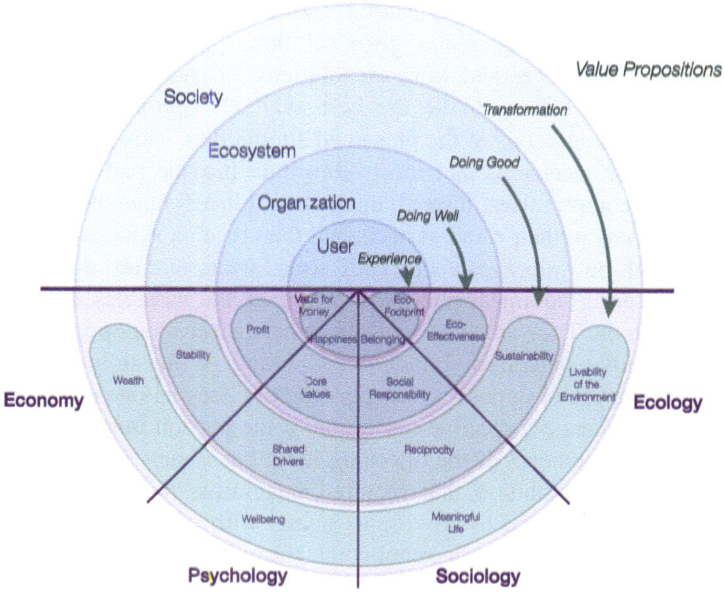

Fig. 5.2 New value propositions for meaningful innovations aim to simultaneously address economical, psychological, sociological and ecological value on all four levels

several stakeholders can join together in an association to explore the possibilities of smart energy grids to drastically improve the sustainability of energy use in cities and villages.

Transformational innovations are different from these examples in that they explicitly address the creation of value at all four levels of the Value Framework (Fig. 5.2).

Innovation Approach

Meaningful innovations aim to realize value: for different stakeholders; at different levels; and from the economic, psychological, sociological and ecological perspective. These different stakeholders in the innovation may have seemingly conflicting needs and requirements, and they may pursue different values. Their conflicting values and needs may be experienced as an 'unsolvable problem', but in practice similar problems have been solved by integrative thinking. This is an approach that enables the creation of a synthesis containing value for all. As Aristotle already emphasized: values are heterogeneous, they can conflict, and man has to make decisions without optimizing rules, making choices means confronting dilemmas [83].

Integrative Thinking As Roger Martin shows in his book 'The Opposable Mind', integrative thinking creates great solutions [51]. Conventional thinkers break problems into pieces and work on solutions for each piece separately, settling for what they believe to be the best available options. They decide by examining the pros and cons of the presumed alternatives. This process inherently leads to 'either-or' decisions that most often only provide partial solutions and make too many compromises. Instead, integrative thinkers have the ability to hold two opposing ideas in their mind at once, and then reach a synthesis that contains elements of both but improves each. They seek factors that are not immediately obvious, considering nonlinear relationships among the variables, and seeing the problem as a whole. They are able to resolve tensions between opposing ideas and generate innovative outcomes. Integrative thinking is the ability to face constructively the tension of opposing ideas, instead of choosing one at the expense of another, and to generate a creative resolution of the tension in the form of a new idea that contains elements of the opposing ideas but is superior to each.

An example of a solution that required integrative thinking is the Citizen M hotels. Citizen M aimed to provide its guests with luxury at affordable price levels [12]. But luxury usually does not come cheap. The Citizen M hotel concept is based on relatively small rooms, with a full-size luxury bed, rain shower and a high-level ambient atmosphere. The first hotel opened at Schiphol Airport in Amsterdam in 2008, and others followed in Amsterdam city center and Glasgow. One of the pillars of the concept is a portable Moodpad, which allows guests to tune the ambient setting to their own personal preferences. They can switch from lighting with warm colors and blinded windows, digital art and music in a cozy atmosphere, to an environment that is more functional and optimized for working. Personal preferences are stored in a central server and on the guest's personal RFID card that also serves as key to the room. The rooms are reserved online, and linked to the personal RFID card, which enables guests to go directly to their reserved rooms without the need to first check-in at reception. The guest's mobile profile, which is linked to the card, also enables preprogramming of the room settings to the personal preferences. The advantage for returning guests is that they do not need to re-enter their personal preferences, but have them immediately available for the next visit, even if this is in one of the other hotels. Next to these advantages for guests, the system also provides advantages for the other users of hotel rooms: the hotel staff. The remote-maintenance system leads to 50% cost reduction, starting by eliminating the check-in and check-out procedures which saves personnel at reception. Cleaning staff can work much more efficiently because the status of the room is automatically communicated to a central office. A remote check can be made of whether all appliances are fully operational, and availability of the rooms can be confirmed. The seemingly conflicting requirements of providing luxury together with low costs have been tackled by bringing together people from different backgrounds. The hotel concept was designed in close collaboration between Citizen M and Philips Innovation Services. One party brought in knowledge and experience of the hotel business and the willingness to choose unconventional solutions, while the other contributed knowledge of all kinds of available technologies and experience to integrate them into innovative solutions.

Gestalt Theory Integrative thinking links to the Gestalt theory. This puts forward the notion that a unified concept is perceived as more than the sum of its parts [60]. This works in the same way as a piece of music in which the melody is still recognizable when it is played in different keys, although none of the notes is the same. If a melody and the notes that comprise it are not independent, then a whole is not simply the sum of its parts, but a synergistic 'whole effect' or Gestalt [4]. The design of the solution is then not just a question of solving smaller problems and integrating them into a whole because the different parts together are not sufficient to explain the whole system [68]. Gestalt distinguishes itself from the 'classical' analytical thinking that starts with analysis and reasoning to come to design decisions. This analytical process requires a deep understanding of the problem. But when looking for holistic solutions to societal challenges, these will typically be 'ill-defined' problems. The nature of ill-defined problems is that they cannot be solved simply by collecting and synthesizing information. All the relevant information cannot be predicted and established in advance of the design activity. The directions that are taken during the exploration of the design territory are influenced by what is learned along the way, and by partial glimpses of what might lie ahead. Intuition plays an important role in this process [14].

Transdisciplinarity Finding solutions for ill-defined societal challenges requires the integration of different knowledge fields. As Max-Neef concluded: none of this century's main problems, such as water, poverty, environmental crises, violence, terrorism and destruction of social fabric can be adequately tackled from the sphere of specific individualist disciplines. This poses a problem, since education systems are mostly monodisciplinary, resulting in experts in rather narrow fields. Although it is frequently attempted, the situation cannot be solved by creating teams of specialists in different areas around a given problem. This team approach only leads to an accumulation of visions emerging from the participating disciplines. But an integrating synthesis is not achieved through an accumulation of different brains. It must occur inside each of the brains, or inside one brain that oversees the whole. What is needed is transdisciplinarity: a more systemic and holistic seeing of the world, as opposed to the partial, fragmented and limited visions and assumptions that stem from the combination of deep knowledge of single disciplines [53], see also Fig. 5.3.

Transdisciplinary vision is transcultural, transnational and encompasses ethics, spirituality and creativity. It is the science and art of discovering bridges between different areas of knowledge and different beings, between humanistic and scientific cultures, between hard and soft knowledge, between natural and social sciences. The process of transdisciplinary problem-solving requires not only scientific knowledge, but also instrumental, ethical and aesthetic forms of knowledge. Stakeholders are entering into a process of negotiation; confronting the four kinds of knowledge in a series of encounters that allow representatives of each type to express their views and proposals. In the process, a fifth type of knowledge

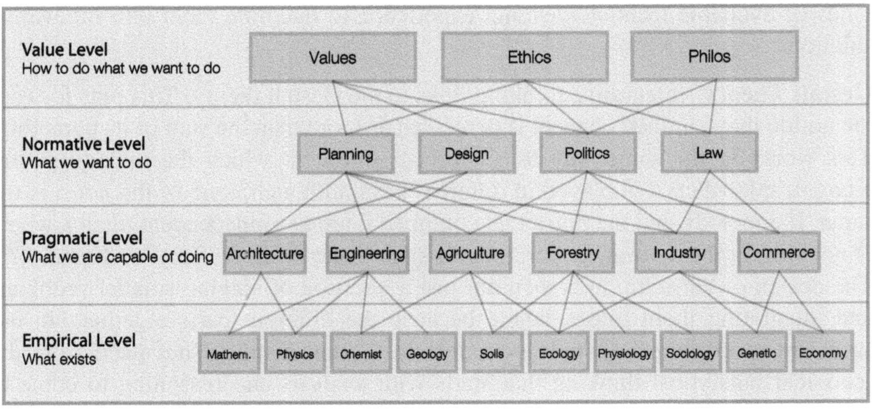

Fig. 5.3 Transdisciplinary action is defined by any multiple vertical relationship covering all four levels [53]

progressively emerges. It is a kind of hybrid product, the result of 'making sense together'. As progressively shared meanings, diagnoses and objectives emerge, individual interests and views are seen in a different perspective [74].

Integrating Views Through Sketching In the early phases of the process, sketches are a good way to represent the ideas and to share them with different stakeholders of the project. Half-formed ideas can be expressed and reflected on: to be considered, revised, developed, rejected and returned to [14].

Sketching supports the process of transdisciplinary problem-solving or 'making sense together' as referred to above, because it visualizes the half-formed ideas, and supports the process of jointly reflecting on them from the different backgrounds and disciplines.

In the Netherlands, Design Initiatief actively pursues social open innovation, by initiating projects and partnerships. One of the activities is organizing kick-off workshops, in which people from profit and nonprofit organizations come together to explore the possibilities for innovative ideas to solve a specific societal issue. During these workshops, ideas are generated in groups of people from completely different backgrounds. Professional designers sketch the ideas put forward by the participants. The sketches help the participants to reflect on the ideas, built on each other's ideas and notice shortcomings that inspire a new wave of ideas. Figure 5.4 shows some examples of sketches that were made during the kick-off workshop to look into novel solutions for consumer healthcare [77].

Wicked Problems The societal challenges that this book focuses on are 'wicked problems': a class of social problems which are ill-formulated, for which the information is confusing, for which there are many clients and decision-makers with conflicting values, and for which the ramifications in the whole system are thoroughly confusing. Wicked problems are ill-defined, and unique in their causes, character and solutions. Dealing with wicked problems demands that special

Fig. 5.4 Examples of sketches design initiatief [77]

attention is paid to understanding the nature of the problem itself. Problem understanding is central; the solution is secondary [50].

Since the problem cannot be fully understood in isolation from consideration of 'the solution', it is natural that solution conjectures should be used as a means of helping to explore and understand the problem formulation. The problem and the solution co-evolve [14]. How many iterations are needed is not known up–front, neither are the problems and opportunities that will be found along the way. Planning of such projects is not a straightforward job.

The challenge of design is then not a matter of getting rid of the emergent, but rather of including it and making it an opportunity for more creative and more adequate solutions. The knowledge to understand, frame and solve problems is not given, but is constructed and evolves during the problem-solving process. Fischer and Giaccardi have identified 'meta-design' as a framework to design for change. Transformational innovations that aim for behavioral changes have strong similarities to meta-design. Table 5.2 shows the main differences between traditional design and meta-design [26].

Meta-design poses another level of wickedness to the design challenge, as it deals with emergence, and adaptive designs. It aims to design systems that are not 'finished' when users start to work with them, and accepts that the users will influence the design of the system by adapting its functionality. Today more and more product-service systems are already designed for change. A well-known example is the Apple iPhone with its apps. When designing the iPhone, Apple could not predict what kind of apps developers would come up with, so it could not envision the complete spectrum of use of the iPhone. Five years after the introduction of the iPhone, new apps are still being developed that take their inspiration from the emerging needs of users, and new ways of using the technology in the phone are constantly being found. A nice example is this respect is the Sleep Cycle alarm clock that uses the iPhone's motion sensor to monitor users' sleep patterns and wake them in the lightest sleep phase. The total system of phone hardware and apps is still evolving.

Table 5.2 Meta-design/design for change as compared with traditional design [26]

Traditional design	Meta-design
Guidelines and rules	Exceptions and negotiations
Representation	Construction
Content	Context
Object	Process
Perspective	Immersion
Certainty	Contingency
Planning	Emergence
Top–down	Bottom–up
Complete system	Seeding
Autonomous creation	Co-creation
Autonomous mind	Distributed mind
Specific solutions	Solutions spaces
Design-as-instrumental	Design-as-adaptive
Accountability, know-what (Rational decisioning)	Affective model, know-how (embodied interaction)

For adaptive systems, and systems that are 'open–ended' like those with new applications that can be added during the product lifetime, the traditional design process that starts with making a user requirement specification is no longer suitable. Writing a comprehensive specification would take many years of work to cover all the possible user scenarios, and even then emerging new user scenarios would not be included. Testing all scenarios is also impossible to do in a reasonable time frame. This means that the classical design process, with milestones at which product designs are released for the following phase, is undermined. Development teams will have to learn to deal with uncertain and incomplete information. Nevertheless, they will still need to ensure that both the technical product quality and the user satisfaction of the systems that are launched onto the market are high. The inherent uncertainty and the need to ensure quality whatever happens impact the planning of such projects [18].

Uncertainty Reduction Through Prototyping Prototypes in any form such as sketches, storyboard, scenarios, movies, scale models, dummies or functional prototypes can support the iterative process of uncertainty reduction. Typically in transformational innovation, uncertainties may be high in all three aspects: desirability, technical feasibility and/or economic viability [8]. This corresponds to the top–front corner in Fig. 5.5.

It is all about learning faster, learning cheaper and learning better. This means that there is nothing wrong with failure, as long as it happens early and becomes a source of learning. In that sense, experimentation is not the same as blind trial and error—it is a process of experimental design and exploration through successive approximations of the products or services. Depending on the project, an uncertainty reduction strategy can be chosen that is supported by a planning of prototypes that supports testing to reduce uncertainties. The prototypes made to test

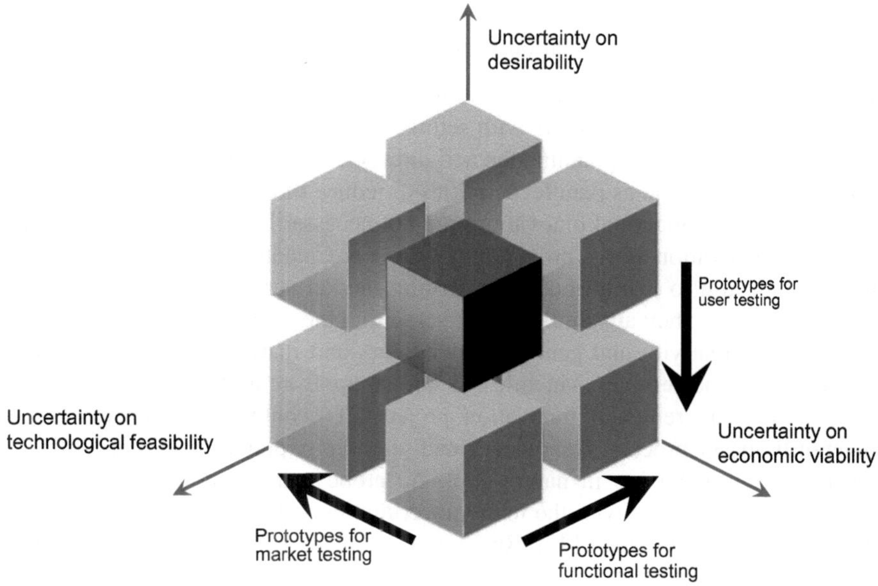

Fig. 5.5 Uncertainty reduction through prototyping

desirability with users can differ from those made to test the technological feasibility through functional testing.

Typically for transformational innovations a more experiential process is required, for example by introducing an early version of the product to an initial market, learning from experience, modifying the product and approach to the market based on that learning, and then trying again, this time with somewhat better information and understanding and somewhat lower uncertainty [47]. However, it is not likely that all the uncertainties will be reduced before market launch. Hence careful consideration should be given to which uncertainties can be minimized during the project, and which ones the team will have to cope with through good feedback mechanisms after the market introduction, to allow for later improvements of the solution.

A common approach to reduce the risks and uncertainties in radical innovation projects is to choose not to address all the issues of uncertainty concurrently [54]. A sequence can be made of the activities that aim to reduce the uncertainty, taking into consideration which uncertainties it is most important to tackle first. In general, understanding the customer looks like a logical first priority [16]. So that means the uncertainty of desirability is dealt with first. As already described, this also includes the option of 'test marketing' the product by launching it early to get direct market feedback. Especially for product-service systems, or systems that are launched as 'open ended' systems to which new applications can be added later, a conscious decision can be taken to go to market early in limited quantities, or to users who can deal with imperfect solutions. These options will be discussed in more detail later in the section on Expeditionary Marketing.

Demonstrator Planning Transformational innovation projects cannot be planned in the traditional way that starts with setting a target launch date, defining the in-between milestones and planning the activities to complete the deliverables for the milestones, preferable as parallel activities to reduce throughput time [16, 69, 76]. Although transformational innovations are not necessarily similar to the definition of radical innovation as e.g. provided by Garcia & Calantone [29] in the sense that they do not aim to disrupt both the existing market structure and existing technology structure, they are similar in the sense that they can rarely be planned. It is through the creativity and genius of innovators and marketeers that they evolve into products and services. Much less is known up–front at the start of the project than in more incremental innovation projects. Discontinuous innovation is an inherently messy process: coincidence and serendipity play an important role, and the process is recursive in nature [80]. A 'probe and learn' approach is more appropriate for dealing with the many uncertainties and complexities surrounding the market and technology [47]. Really new products will likely involve iterative learning from the market as customer requirements and technological capabilities co-evolve [71]. This is because market opportunities tend to be unspecified and first need to be created, and because potential users are often unable to envision the true potential of a really radical innovation [80].

But this does not mean that planning is not important. In most cases the participating organizations include for-profit companies, and some of them have the good practice of aligning their market launch activities with big events in the market. This means that they will aspire to be able to launch an innovative solution at a certain date to increase its visibility and facilitate fast adoption of the solution. For example, the big Consumer Electronics Show (CES) in Las Vegas, held annually in January, attracts retailers, customers and journalists. Products launched or announced at the CES are certain to have a large audience and lots of media attention. This is an opportunity, but a risk at the same time. A straight 'first-time-right' process cannot be defined for transformational innovations. Because the problem is ill-defined, and because of the characteristics of wicked problems, knowledge of multiple disciplines is required to even start to understand the problem or opportunity. The process has an explorative nature [36]: along the way new discoveries are made that either lead to upgrading the concept or rethinking the assumptions [8]. Sometimes ever more interesting or promising opportunities arise, and it would be foolish to ignore these. This leads to a highly iterative process. This is also a reflective practice: experimental actions are taken and the results scrutinized, leading to a better understanding of the situation [78].

An approach for planning radical innovation projects that has proved its value in practice is the planning of demonstrators. Similar approaches like 'street-testing' [57] or 'discovery driven planning' [36] are described in the literature to install fast learning cycles to support decision-making in innovation projects. In this approach the project activities are driven by uncertainty reduction and

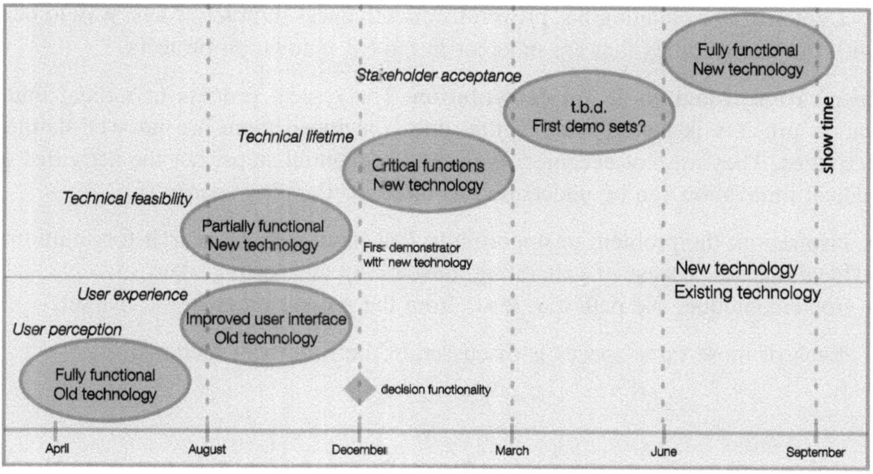

Fig. 5.6 Example of demonstrator planning

knowledge gathering. A timeline for the project is plotted, showing different important dates for the project. First of all, important external events such as trade shows and conferences are shown. These events can be used for two different purposes: to gather information that is relevant for the project, but more importantly also to get feedback on the project by showing demonstrators to the public, potential business partners or other stakeholders.

Secondly, the timeline is used to indicate the deadlines for demonstrators that are used to reduce the uncertainty. A sequence of demonstrators are planned to check the desirability, feasibility and viability of the project. This planning helps the team decide which uncertainties need to be tackled first, and what are the requirements of the demonstrators that need to be produced. The shorter the iterations the better, so it is better to take more and smaller steps than one giant step.

An example of a demonstrator planning is given in Fig. 5.6. In this project it was decided to first explore the desirability of the value proposition with the prospective users by showing the benefits using the existing technology. The new technology would deliver additional features in the user interface at a much lower price level. But it was not clear if this would offer sufficient benefits to the intended users to make them switch to the new solution. After completion of the user tests, the planning continues with the technical tests that also cover lifetime testing of the new technology. In making the planning the team realized that they would need to do additional tests with project stakeholders such as the retailers, to understand their acceptance of the new solutions. This is included in the planning, although without all the details of the demonstrators which the team was not yet able to define. When uncertainties like these are encountered it is recommended to show them on the timeline as placeholders, to at least act as a reminder that specific actions still need to be taken to provide clarification. A demonstrator planning is not static; it needs to be updated with advancing insights.

Demonstrator planning has proven its effectiveness in practice as a way to deal with the uncertainties that are inherent in radical innovation projects.

Inspiration, Ideation & Implementation The typical process of radical innovation projects is not linear, but that does not mean there are no well-defined activities. They are not executed as orderly sequential steps, but the activities of radical innovation can be understood as three overlapping spaces [8]:

- Inspiration: the problem or opportunity that motivates the search for solutions;
- Ideation: the process of generating, developing and testing ideas;
- Implementation: the path that leads from the project room to the market.

Each of those three spaces is discussed in the following sections.

Inspiration

The first space in innovation is to identify the problem or desire for which a new value proposition is to be defined, and that motivates the search for solutions. As described previously, the inspiration for meaningful innovations can be found in any of the segments of the Value Framework, as shown in Fig. 5.1. Current problems can inspire new ideas, as can opportunities found in trends or visions of the future.

Understanding the Challenge Regardless of where the inspiration was found, the first step is to explore the challenge. This involves gathering information on the specific problem or opportunity. This information reflects both the current situation and visions of how it could be in the future.

For example, Design Initiatief in the Netherlands investigated possible future scenarios for the Netherlands in 2024 [66]. Based on discussions with opinion leaders from various sectors, they identified social, technological, economic, ecological and political developments. Different scenarios for the Netherlands in 2024 were explored through serious gaming, creating a wide range of possible futures full of changes that inspire new business opportunities. These futures explore trends like hyperconnected communities, magical technologies and power to the people. The project resulted in six opportunity areas for sustainable value innovation: the vital human, the future of food, the use of energy, the Netherlands as a knowledge nation, water and the safe society. Workshops were organized in each of these areas to explore them from multiple perspectives, and to understand the challenges at a more practical level.

Once the challenge is more or less clear, the next step is to understand the needs and desires of the intended customers of the new value proposition.

For Whom is Value Created? The first element of a value proposition is for whom it is intended. In the literature this is often translated immediately into the customer, who is also the buyer of the solution. But this is not automatically the

Table 5.3 The different potential customers for a new value proposition

Customer	
Users	Indication of unmet needs and aspirations of the separate relevant user groups, and where relevant the differences in capabilities and expectations
Buyers	Indication of the motivational drivers of the people or organizations buying the solution
Influencers	Indication of the relevant influencers of the buy or use process and their main interests

case for the innovations in this context. For example, the users of medical equipment for hospitals are both patients and medical staff. However they are not the buyer of the solution, which in this case is the hospital. There are many examples of products and services that are not bought directly by the users. In the Netherlands, customers who subscribe to digital TV services get their set-top boxes from the service providers, and themselves have no influence on the brand or type of set-top box they use.

A distinction will have to be made between the different stakeholders of the value proposition: the users, the buyers and the influencers, see also Table 5.3. The stakeholders who contribute to creating the solution are not yet included at this stage, and will be discussed later in Chap. 6 on designing new ecosystems. The focus here is on the stakeholders who are the intended customers for the solution.

Users First of all there are the users of the solution. They may have to be divided into one or more groups, depending on the situation. If there are relevant differences between different user groups it is important to address them separately because they may have different needs, aspirations, abilities or expectations. Next to the example of patients and medical staff as different users of a medical system, it is clear that specific user groups have different capabilities. So just addressing 'the elderly' as a population might not justify the difference in abilities of the elderly, depending on their health and well-being. The users are considered the most important customers of the value proposition for transformational innovations. Without users who really adopt the product or service for continued use, the proposition will not achieve its intended effect. This means the unmet needs and aspirations of each of the relevant user groups have to be clearly understood.

Buyers The second group of stakeholders is the buyers of the solution. They are the people who will need to provide the money or other value investment to acquire the solution, either for their own use or for use by others. Understanding their motivational drivers is crucial to be able to assess if they will be willing to buy the proposed solution. Here again it is important to consider the differences between buyers, and if necessary to separate them. So in the case of solutions for the elderly, the buyers may be the elderly themselves, their insurance companies, their family members or the care homes in which they live. Each of those potential buyers will have different motivational drivers to buy the solution. While a care home may look into the time the solution will save for staff, family members may find quality of life and the assurance of knowing that their family member is being well taken care of more important.

Influencers Last but not least there is the group of influencers. In some cases there may be people or organizations that have a strong influence on the decision to buy and/or use the solution. Examples of such influencers can be patient and consumer associations, interest groups, religious institutes, intermediates, family members etc. The social contexts of the user and/or the buyer need to be understood to assess the impact of such influencers on the decision process.

What is the Need or the Aspiration? When looking for value propositions in a 'white space' it is not sufficient to try to figure out what kind of products or services people are trying to buy. Instead it is important to understand what they are trying to get done in their lives in given situations [39]. Identifying a need or an aspiration requires adopting a proactive, outside-in approach, as opposed to a passive, inside-out approach (such as looking at the products or services that are currently selling well).

Identifying the needs or aspirations is never a straightforward process. People have complex, multiple needs and aspirations which are not always able to articulate. The difficulty with more radical innovations is that by their very nature they are difficult for customers to understand and appreciate, due to the lack of logical product evolution. Business analysis and market research should be conducted as soon as the relevant questions and an appropriate means for answering them can be determined [80]. In practice this means that the traditional market methods cannot be used in the initial stage of more radical innovation [16, 72].

Insight Insight, or learning from the lives of others, is one of the key sources. Insight usually does not come from quantitative data; a better starting point is to go out into the world and observe the actual experiences of people as they improvise their way through their daily lives, and do the 'thoughtless acts' that they are accustomed to doing. People are rarely aware of what they are doing, and what to do to improve it, but their actual behavior can provide invaluable clues to their range of unmet needs. The advantages of searching for insights are that they can be found everywhere, and they are free [8].

Ethnographic research is one of the common approaches to gather insights. Ethnography provides insight into the organization of social settings, models for thinking about those settings and the activities that go on there [23]. Ethnographic research involves going into the field and describing what relevant persons do in the specific situations, and explicating what they have to know to do the activities, and how that knowledge is deployed in the ordering and organization of the activities [9]. Ethnography has always been very appealing to designers because it investigates not just what users say, but what they actually do [82]. How to conduct ethnographic research will not be described here in detail, as sufficient literature is available on this subject.

Different approaches have been suggested in the literature to support the process of identifying unmet customer needs, such as observation of customers

[25], 'extreme' users [8] or lead users [81]. Enough time must be spent to reveal both 'normal' and 'abnormal' conditions of use, which makes it an expensive approach, and even if it is unobtrusive, it can change people's behavior, so 'natural' actions may not be captured [33].

One thing to take into account, specifically when working on transformational innovations, is the fact that people can frame their choices differently depending on the roles they see for themselves. For example, people might give different answers depending on whether they are reflecting from their own utility as users, or as citizens on behalf of society [6].

The best way to gain deep insights is to involve users in the process; to make them co-researchers. Together with the team they analyze their own lives, which enables an inside look at their lives, supplementing the outside view of the team. Not only is the user's home or work context taken into account, but also the total context in which he or she does things, focusing on the combination of individual time, place and activity. In this way, an understanding is developed of what people consider to be relevant, meaningful and understandable [46]. Many societal innovation programs involve users, such as the projects of the Young Foundation in the UK, or the Livable Cities program of Philips.

Empathy 'Standing in the shoes of others.' All field notes, videos and photographs are useless unless they are translated into insights and these insights into products and services that will improve people's lives. The purpose is not to generate new knowledge, to test a theory or to validate a hypothesis. But it is all about empathy: a mental habit of making the effort to see the world through the eyes of others, understand the world through their experiences and feel the world through their emotions. An emotional level of understanding of the people in the target group is needed. What do they feel? What touches them? What motivates them? [8].

But empathy for the individual by itself is not enough. Today an understanding of the social interactions of people within groups and the interactions between groups is also required. This is especially true for web-based services, but as more and more people gather information and have social interactions online it is also true for other products and services. In some cases people even create applications or user-generated content, and become part of the innovation process themselves [62]. The boundaries of creators and consumers are becoming blurred at the level of companies and individuals.

Once insights are obtained into customers' needs and desires, these need to be captured in a consise,contextually specific customer statement in the words of the customer [33].

The value of the new solution, in other words the potential success of the new value proposition, is defined by the importance of the need or desire to customers and how satisfied they are with the current solution. Once the insights are captured, the next step is to frame the problem.

Problem Framing Problem framing is reformulating problems and making them explicit. It is an important action that defines the success of a team [78]. Especially ill-structured problems require a framing action [22]. Making ill-structured or wicked problems explicit and developing their characteristics is not a clearly defined task, but is still an important part of the project. It involves selectively viewing the situation in a particular way. Parallel lines of thought are often used to explore the viewpoints of different stakeholders or different aspects of the problem [45].

The framing process includes setting its boundaries, selecting particular things and relations for attention, and imposing on the situation a coherence that guides subsequent moves. A statement is made that summarizes how the problem is seen or the structure of the situation that the problem presents. This problem-structuring activity is not only done at the beginning of the project, but also recurs periodically [30]. Problem framing is an important action in meaningful innovation. It supports the process of 'making sense together'.

Personas Personas have proven to be a useful tool in the creation of meaningful innovations. They are a precise description of users and what they wish to accomplish. They are not real people, but hypothetical archetypes of actual users. The relevant personas and their goals are determined during the investigation of a problem domain. Cooper [13] provides a detailed description of personas and how to create them, as do others who implemented the method in their design projects [7, 63]. Although personas were initially intended to support the process of designing user interfaces of products, the approach has also been proved in practice for more holistic purposes [61].

How many personas are needed to represent the important stakeholders depends on the project. When exploring a new field, more personas are usually needed to understand the different perspectives of relevant stakeholders. Not necessarily will all these stakeholders be part of the target group of the new solution, but a multi-stakeholder perspective is needed to understand the context of the problem, and to find the deeper needs of various people that are related to the societal problem or opportunity at hand. Descriptions of personas vary with the project. They need to be tuned to capture the relevant information for the specific problem or opportunity. Thus for healthcare-related projects, it is important to include both medical staff and patients, and also patients' coping strategies.

Experience Flow Chart Personas are an important tool in understanding the specifics of the different stakeholders, but they lack a time perspective. As meaningful innovations also aim to address transformations, it is important to get a wider view. Personas and ethnographic research provide insight into important moments, but observing patients who are scanned in a hospital only provides a partial view of the total experience that they go through. Understanding people's experiences and needs demands ways to also capture their environmental context and conditions [86]. These may be found in the home environment, public spaces, rehabilitation centers, private clinics or the general practitioner's consulting room in the case of healthcare. Plotting customer journeys have proven to be very helpful in understanding people's experiences. For example, to understand the experiences of onset dementia patients an experience flowchart was created to visually represent their journey (Fig. 5.7). Three personas were created to represent specific characteristics of patients (e.g. a patient who is not accepting the

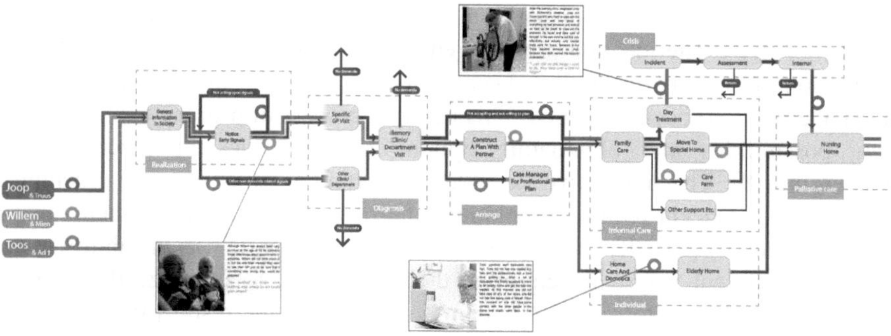

Fig. 5.7 Experience flowchart for dementia patients with three personas and examples of their experiences [1]

diagnosis, a patient who is willing to cooperate and make the best of it, and a person who is unaware of her condition). In an interactive flowchart videoclips, photos and stories can be used to further illustrate the specific experiences, emotions and unfulfilled needs of these patients and their loved ones [1].

Such a flowchart allows the team to walk through the healthcare journey as seen through the eyes of multiple stakeholders: patients, loved ones and clinical care teams.

The flowchart supports the discussion in the team, and helps to challenge and enrich the insights. It has proved to be successful in making sure that everybody involved is aligned with what they know and what they realize ,they do not know or need to know. It supports the exchange of knowledge between people from different backgrounds, and helps to complete the picture of understanding together [61].

With the help of the experience flowchart, a team can jointly discuss which insights are new, which areas of the experience flow are underappreciated or in which stages of the journey there are increased stress levels that could be tackled. This provides triggers to inspire the ideation process to find new solutions.

Ideation

The second space in meaningful innovation is ideation—the process of creating ideas for new solutions. This is a highly creative process, and can be supported by many tools which will not be discussed here. What is important in ideation for meaningful innovations is that the ideas are generated at a sufficiently high level. If the different (partial) problems and needs identified in the observation of stakeholders are directly taken as goals, the solutions will also be partial. The experience flow method will help to pinpoint important insights from the total experience journey by the key primary stakeholders, and these insights can be used to inspire an ideation session.

As already described, inspiration, ideation and implementation are not sequential steps. The ideation process generates solutions that will inspire further discussion of the experience flow, because ideas can trigger discussion of insights that have not (yet) been identified. This iterative process of probing the future with new possibilities and eliminating options requires constant diverging and converging [8]. The integrative skills are important for this process: reaching a more holistic concept requires combining and recombining different inputs into a structure, thereby partially redefining the input. Also important is the ability to see things in different ways, to determine their meanings, to organize them into a structured whole and to reorganize them depending on the result [44].

The starting point for ideation is an ill-defined problem, even if an extensive experience flow is available. Experience flows provide good insights, but they do not provide exact details that can drive an engineering design-like process. An analytical approach that decomposes the problem, finds solutions to the sub-problems and combines these partial solutions into a total optimal problem solution is neither possible nor desirable. Moreover, during the initial phases of exploring the ill-defined problem the problem itself is redefined using the advancing insights [30]. It is much more realistic to follow the strategy of finding a satisfactory solution, rather than expecting to be able to find an optimum solution to a well-defined problem [14].

Creating Alternatives Many books and articles about design describe the design process as if it is carried out by a single person. This will hardly ever be the case in finding solutions for societal challenges. Different viewpoints are needed to bring together the knowledge and experience of the users, their needs and aspirations on the one hand and the possibilities in technologies and solutions on the other hand. This means that the process is much more a co-creating and co-evolving process that brings together different disciplines to jointly develop insights and solutions.

The creation of alternatives has always been a success factor for innovative design. Design teams that have explored multiple solutions tend to have made a more comprehensive assessment and have a better understanding of the problem [14]. It is important for the team not to get fixated on a particular problem too early, as this will stop the process of exploring the problem in a more holistic way, and the resulting solution will often be only a partial solution.

On the other hand, if the team sets no boundaries at all for the process they are likely to get lost in the many possible ideas and solutions and have no idea of how to filter out the promising ones. Everyone who has participated in a brainstorming session in which the starting question was broadly defined knows that it is easy for even an only slightly creative group to come up with hundreds of ideas in one hour. The problem then shifts to how to figure out which of those hundreds of ideas make sense in this context, and to use a process of idea selection that does not discard valuable ideas, but nevertheless helps to converge. This is a difficult task if the boundaries are not set clearly. The risk is that people will become demotivated because their ideas get lost.

Thus the trick in exploring alternatives is on the one hand not to start with too narrow a definition and to jump to (partial) solution, but on the other hand not to set the boundaries of the search space too broadly. A balanced search for alternative solutions is crucial. First solutions, even during the process of understanding the problem, should not be suppressed as they can help to gain that understanding, but the team should return repeatedly to clarifying the problem. New variants need to be produced continuously in that process, but to keep the number of suitable variants manageable they need to be assessed consciously to pick out the interesting ideas [27].

The willingness of a team to reconsider the early concepts during the process has proved to influence the quality of the resulting design [70]. This reconsideration can relate to the complete concept or part of it. It may be necessary here to reconsider the concepts because of problems that are encountered in the design process, or because new insights have been gained. One way to fast-track the gaining of new insights is through experimentation.

Experimentation Experimentation is an important activity in the design process of meaningful innovations. As described earlier, prototypes of any kind are useful to explore the field and reduce the uncertainties in the project.

But experimentation is also an important activity in generating solutions. As Kelley puts it: prototyping is 'thinking with your hands' [41]. Prototyping allows the exploration of many ideas in parallel [8]. Cheap and easy materials like cardboard, foam and existing objects can be glued together to demonstrate the concept.

The demonstrators can be used for discussions with users and to get feedback on the concept. This usually inspires a wave of new ideas. As new ideas arise that enrich the concept, they can be added to the existing concept. In the design process used in the Philips InnoHub, an explicit step is included to enrich the consumer experience of the concepts with ideas, as shown in Fig. 5.8.

Demonstrators are useful for the team itself, and to check the concepts with users, but they can also be very helpful for discussions with other stakeholders. For example in its 'Connected Care' project, Philips Design created a vision for a design solution that helps to provide an easy start to lifestyle changes affecting the daily routines of patients. The vision was used to create scenarios to study how the solution can be integrated into people's lifestyles. These scenarios were discussed with users in the early stages to test whether the ideas were going in the right direction. But to communicate the vision to a wider range of stakeholders, working demonstrators were created and presented in an exhibition with live demos, and used in movies and brochures [5]. This provided valuable feedback on the vision from multiple perspectives.

Getting Feedback on Transformational Innovations The aim of meaningful innovations is to address unmet needs with unique solutions. These innovations aim to fill a 'white space'. To understand whether the proposed solution actually provides the expected benefits to the user, it is common to perform benchmarking to compare it with alternative solutions. When dealing with incremental innovations the solution is usually more or less similar to what is already offered on the

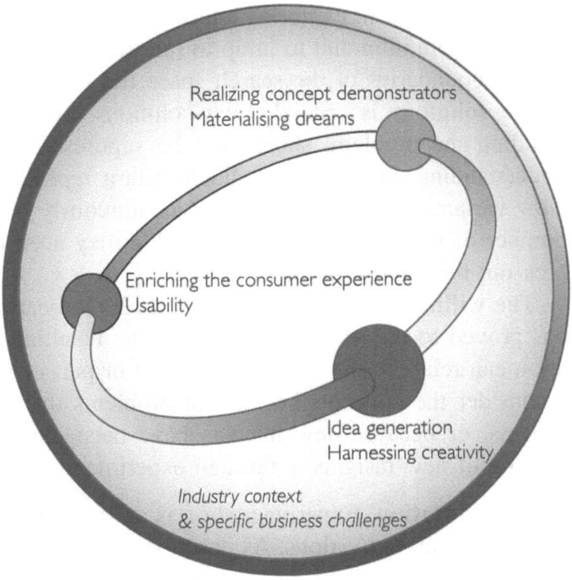

Fig. 5.8 Philips InnoHub process—Going the full circle [19]

market, with some additional functions or features. More radical innovations can be completely new, and even disruptive. But for radical innovations, no similar solutions are yet available. This does not mean that benchmarking is not important, as prospective customers will always compare the solution with alternatives. One alternative that they will surely consider is to accept the situation as it is, and adopt a new solution.

Unique solutions can create 'blue oceans'. These are defined as untapped market spaces that create a demand. Competition is deemed irrelevant in blue oceans, because the rules of the game are waiting to be set. From the viewpoint of an organization this is an important aspect, as it provides the opportunity for highly profitable growth [42]. But it is less relevant from the viewpoint of the customer, for whom the convincing benefits of the new solution over existing solutions are the most important thing.

Getting valuable feedback on transformational innovations is a challenge. Often users, and other relevant stakeholders are not able to imagine the solution to its full extend. Project teams then need to be creative to find ways to demonstrate the solution in a believable way to get in-depth discussions with the users and stakeholders. To give an example, The LUXLAB lighting design agency is asked to design a solution for a cycle path along an ecological zone in the Netherlands. The proposed solution uses dynamic control of different lighting settings at different times to accommodate both the cyclists in their feelings of comfort and safety as well as the wild life in the eco-zone in their quality of life. In the early evening visual comfort can be increased as people need more light when dusk is setting in than later in the night when their eyes have accommodated to the lower light levels. Then as traffic eases, the light will dim to a monochromatic greenish light that is less disturbing for the animals and plants, but still provides good visibility for cyclists. Later, when there are hardly any cyclists using the path, the light is further dimmed to the equivalent of 'full moonlight' provides more natural lighting that is aesthetically more pleasant and requiring even less energy. In case of an emergency the system automatically gears up to increased lighting levels to ensure maximum safety. In the morning a bright cool white lighting setting is used to increase the alertness of the cyclists. This solution is different from traditional lighting solutions as it aims not just to reduce energy use, but to do it in a way that also increases the quality of life in the ecological zone and safety of road users. To get feedback on this solution a demonstrator was created which was shown during the 'Liberation of Light' exhibition, and feedback of visitors to the exhibition was asked. Next to this feedback, specific workshops with different stakeholders were conducted, such as the municipality, neighbors, police as well as different users: school children, athletes who use the path for their weekly running exercise and elderly. In the demonstrator realistic light levels were demonstrated in darkened corridors, and this allowed people to really experience the lighting levels and assess the concept. In the workshop the feedback was collected and a more elaborate discussion was held on the validity of intelligent lighting solutions in the surroundings of an ecological zone.

When comparing solutions, it is important to compare the holistic view. For example comparing an MP3 player with an Apple iPod is not just about comparing the products, but the total experience as perceived by the users, which in this case also includes buying music and transferring it to the player. So the total product service system of the iPod with iTunes should also be taken into account in the comparison. Many innovations are not necessarily based on providing new functions or features, but on providing an easier, more effective or more affordable way to complete a task.

Describing an Initial Value Proposition The inspiration space defines the ideas about the needs or aspirations of users and customers, and the ideation space explores the solutions. Both provide input for the description of an initial value proposition. A value proposition is an important tool, as it enables alignment of all the different people who are working on the project by capturing the core elements in a coherent description that can also be used in reflection with the different stakeholders. Table 5.4 summarizes the main elements of a value proposition [39, 67, 72].

Value propositions are these multifaceted bundles of product, service, price, communication and interaction that customers experience in relationship with the supplier. It is the customer's perception of the proposition that is important, not the supplier's. A supplier may believe it has a service advantage; but if the customer fails to agree then this is not a source of value [84]. Value propositions address the deep truth about what customers really value and how the service or product offering will satisfy those needs [72].

Initial value propositions starts from the perspective of the users. They are the primary target group of any innovation, and without them adopting the innovation the innovation cannot be claimed successful.

Table 5.4 Main elements of a value proposition

Element	
For whom is value created?	Who are the primary users? Who are the buyers (and if applicable the influencers)?
What are the needs or aspirations?	What are the insights on the target group? What specific needs or aspirations do they have that are currently not (sufficiently) met? What is the deep truth about what customers really value?
What is offered as a solution?	What product, service or combination of those is offered? How will it seduce the users and satisfy their needs?
What alternatives are available?	What other options are available (including accepting the situation as it is)
What are the differentiators?	Why would they choose this solution over alternatives? What are the benefits? How is the offering superior?

Fig. 5.9 Advertisement picture of the Philips wake-up light

To give an example of a value proposition: the Philips Wake-up Light (Fig. 5.9) is targeted at men and women aged between 25 and 55, with average or above-average income. They experience difficulty in waking up, and often do not have enough energy to get started on their daily activities. An energetic feeling is certainly desirable as they have to go to work early and have lots of other activities to do or, are very busy taking care of the family. They are open to new (technological) solutions. The insight of this target group is, in the words of the target users:

> In the autumn and winter months I wake up feeling as if it's still in the middle of the night. I want to go back to sleep for at least another hour, otherwise I feel lethargic and a bit low. In the summer, it's so much easier to wake up with more energy; I can start doing things right away. It would be so nice if I could always wake up easily and with as much energy as I do in summer!

The solution that is offered is a wake-up light that simulates the rising sun in your bedroom and gently prepares your body to wake up. The light falls on your eyes, sending your brain a message to reduce the production of melatonin, the sleep-inducing hormone, and to increase the production of cortisol, the energy hormone. You can also choose to wake up to one of three natural sounds that gradually increase, just like the sounds of summer morning.

Alternative solutions that are available are basically of three different types. First of all there are traditional alarm clocks and clock radios that focus on waking up in time, a gentle process is of secondary importance. Sony and Philips Consumer Lifestyle are the largest players in premium alarm clocks. As substitutes, people use coffee and tea or even take a refreshing shower. A third category of alternatives is 'dawn simulators', a category in which no consumer A-brands are present (only small online sales). The benefits of the Philips Wake-up Light are that it simulates dawn in summer to wake you up. Now you wake up as easily, naturally and with as much energy as in summer.

As the example shows, a value proposition can be formulated briefly (usually no more than one A4 page). But it provides the key information needed to discuss and challenge the proposition, and to align the team working on it. A well- formulated value proposition can be discussed with users to test whether it indeed appeals to them, and can be discussed with important stakeholders or potential partners to check their interest and commitment to the project.

Enriching Value Propositions The Value Framework can be used to enrich initial value propositions. This allows more stakeholders to be addressed, and more shared value to be created. The process will be illustrated using the example of Philips Ambient Experience hospital labs.

The project started at Philips Design, the division that is responsible for the design of Philips products including medical systems. Instead of taking the traditional approach of implementing new technologies to improve the quality of the medical images created during scans with magnetic resonance or X-ray computed tomography, a different approach was taken, starting with research into the personal experiences of patients and medical staff. This was followed by a close look at the phases patients have to go through during radiological examinations, from expectation to memory. The needs of patients and professionals during waiting, preparation and the actual scanning procedures were identified and translated into solutions. This led to the radical conclusion that the whole radiology department should be redesigned. In fact an entire architectural solution needed to be designed to design an experience that makes patients feel comfortable. Philips Healthcare had to change its proposition from supplying 'boxes' to being an organization committed to working in partnership with the construction

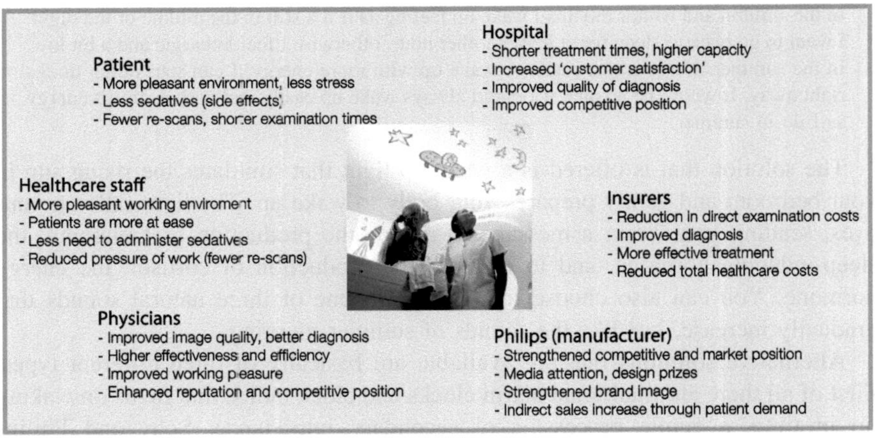

Fig. 5.10 The value of an ambient experience hospital lab—examples of value for stakeholders

industry, architects, social and cultural researchers, human behavioral psychologists and all the other disciplines required to conceive a whole hospital environment [43].

The research revealed that one big problem in making medical scans is that patients are often agitated, and do not lie still. This applies to adults and children alike. In practice it means that images are not sharp, and it is harder for the physician to make a diagnosis. If the quality is too low, patients have to be scanned for a second or even third time. Sometimes patients become so agitated that they have to be given a sedative to allow the scan to be made. This is not without risk, especially with children. Traditionally, Philips had been working on technologies to allow scans to be made more quickly. A side effect of the increased speed is that patient's movements have less impact on the image quality. But when it became clear that the movements actually had such a great effect, another solution that came up was to reduce patients' agitation.

In the Ambient Experience hospital labs, technology like lighting, film projection and sound are used to transform the lab into a space in which patients feel at ease. They can choose a theme, and the film projection and sound distracts them so they lie quietly. Special films for children help them to hold their breath for 20 s at just the right moment. As this example shows, putting the end-user at the center of the innovation process leads to other solutions than those based on a technology perspective.

But such solutions also require a broader context to really illustrate their value. Figure 5.10 shows some examples of value for different stakeholder. For the hospital this solution provides additional value over the traditional imaging equipment because it reduces the scanning time per patient, resulting in higher throughput. Staff and equipment are used more efficiently, so the overall costs are also reduced. But the higher quality of images also allows physicians to make better diagnoses, which in turn leads to more effective treatment. Nurses not only

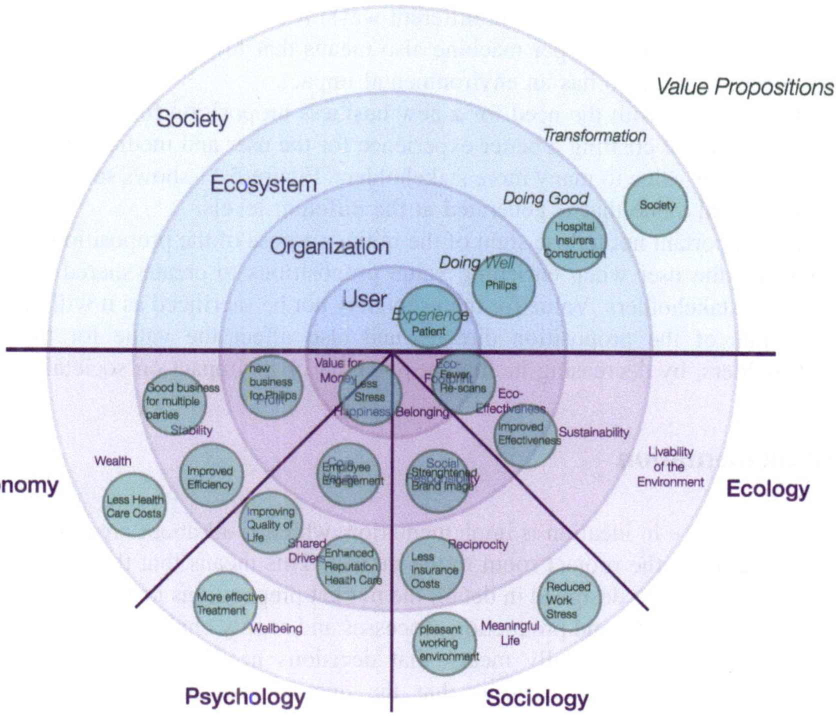

Fig. 5.11 Enriching a value proposition: example Philips ambient experience hospital lab

have a much more pleasant physical working environment, but they also get to work with patients who are less stressed, and they are relieved of having to give patients sedatives. With patients having more freedom to choose between hospitals, and the increasing number of patients who explore their options on the Internet, positive patient experiences that are posted online can increase the attractiveness of a hospital and strengthen its competitive position. For Philips, as the manufacturer of the product, media attention and improved brand image can even result in patients approaching the hospital and asking if they use the Ambient Experience solution.

For health insurance companies the added value lies in the reduced costs of each examination, and indirectly in the reduced costs of the resulting more effective treatment. This leads to an overall reduction in healthcare costs, which in turn also benefits society as a whole. Society itself benefits from better treatment, and improved quality of life of patients. The effects on children can be particularly important for society: less traumatic experiences can impact a child's well-being, and moreover the risk of side-effects through sedation is much higher with children.

Indirectly there are also positive effects on the planet. Higher efficiency and effectiveness of the process mean that more people can be helped with less equipment. This saves energy in different ways: fewer scans per patient are direct gains, but more patients per machine also means that fewer machines need to be produced, which also has an environmental impact.

What started with the need for a new business proposition for Philips, and the limited scope of creating a better experience for the user and medical staff, turned out to deliver value to many more stakeholders. Figure 5.11 shows some examples of additional value that is generated at the different levels.

It is important not to lose sight of the initial purpose of the proposition to create value for the user when enriching value propositions to create shared value for multiple stakeholders. Value for the users may not be sacrificed as it will decrease the value of the proposition directly and also affect the value for the other stakeholders, by decreasing its market potential or its impact on societal level.

Implementation

The third space in ideation is implementation, which is all about creating the path that leads from the project room to the market. This means that the products and services need to be designed in detail, the market preparations need to be made for selling the products, and production processes and supply chains need to be set up.

Implementation basically means that decisions need to be taken about the market launch. One scenario is that the uncertainties have been sufficiently reduced to take the bold decision to launch the product or service fully: possibly worldwide in large quantities. To ensure that such a project succeeds, it has to follow the more traditional working approach with milestones after the concept direction is chosen. Several descriptions of such milestones and stage-gate processes are available, and these will not be discussed here in detail [69, 76].

Figure 5.12 shows an example of a process that describes the deliverables for different aspects such as finance, marketing, product, production system, production process, supply chain and organization for a product that needs to be mass produced. Such checklists for deliverables help a team to ensure that the important activities are planned in time, and are not overlooked. Of course each project is different, and a check needs to be made of whether specific requirements have to be met. For example, healthcare products need to be approved by the Food and Drug Administration (FDA).

For transformational innovations, the uncertainties are unlikely to be low enough to plan a full-scale market introduction. Most often the uncertainties that were part of the project when the initial value proposition was defined are still so high that an incremental working process is not suitable. In these cases another scenario is a limited market launch. This can be decided for different reasons: to test whether users are indeed willing to adopt the product, or because the technological uncertainties are too high. The limited quantities also allow after-sales service to be provided to a limited number of customers to further improve the product or service.

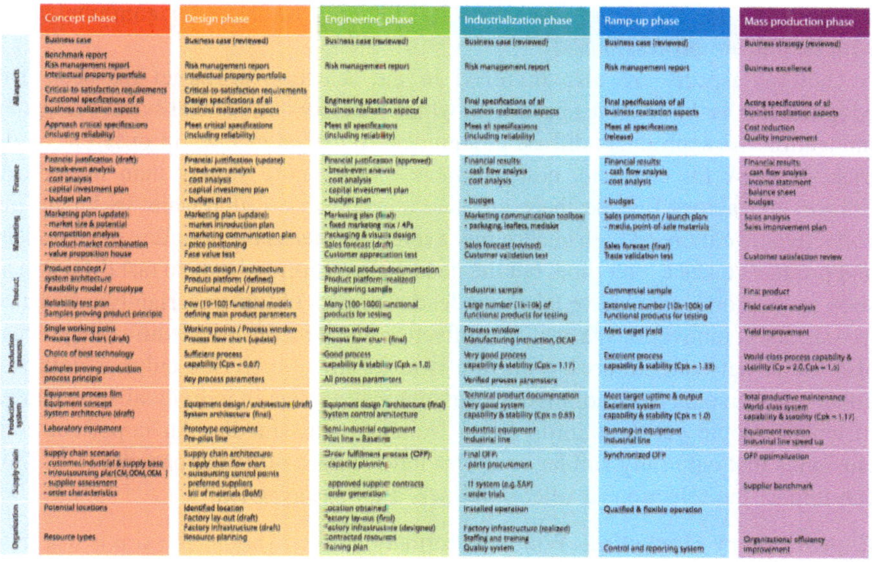

Fig. 5.12 The Stargate process of Philips innovation services

Target Market Selection In choosing the market in which to launch a radical innovation, it is important to consider the openness to adopt new things among the target group. People differ in their openness to adopt new solutions on an individual basis. As Rogers has pointed out, many people perceive an innovation as uncertain or even risky, and hesitate to buy or use it. They will look for others, like themselves, who have already adopted the new idea. The diffusion of innovation thus starts with a small number of people adopting the idea, and then spreading the word to others. Consumers can be divided into five adopter categories: innovators, early adopters, early majority, late majority and laggards [65]. People can belong to different categories, dependent on their interests: a person may be an innovator for audio equipment, but a laggard for domestic appliances.

Understanding these categories is important when selecting a target group to which an innovation is presented at market launch. Certain solutions will need to be directed at communities that are less open to adoption of new things. This may require careful preparation of the market. For example, Maat Foundation in the Netherlands is currently rolling out a video communication platform in certain neighborhoods with larger numbers of people requiring medical care. Ultimately their aim is to have these people connected to two-way video connections with their general practitioner and medical specialists. However, this target group is not used to such technologies, so a low-profile value proposition is first being offered to allow them to get used to the technologies. The expectation is that these users will have an easier transition to the new solutions if they are taken along in small incremental steps. In the meantime Maat Foundation is also learning from the experience, and can use the feedback it gains to improve the solution and make it user-friendly for the target group before switching to using for communication with healthcare professionals.

Cultural aspects are also involved in the selection of target markets. The adoption rate is the relative speed with which members of a social system adopt an innovation. New products diffuse at significantly different rates in different countries. In USA for example, new products diffuse more slowly than in Asia or Europe, but across Europe there are also considerable differences. In Japan, the collectivistic culture of high uncertainty avoidance means the adoption of new ideas and products takes a long time, but the need for conformity leads to fast diffusion as soon as opinion leaders have taken the lead. Cultures of low uncertainty avoidance in Asia adopt innovations faster. The influence of the media on the diffusion process also varies between cultures. Diffusion is faster in collectivistic cultures with frequent interpersonal (word of mouth) communication than in individualistic cultures in which people get their information from the mass media. An important conclusion is that culturally similar countries have similar diffusion patterns [17]. This can impact the selection of where products are first launched.

Expeditionary Marketing With transformational innovations the aim is to create new business with new solutions. For such innovations it is usually impossible to know in advance what configuration or product and/or service features, offered at what price point, through which channels will be required to unlock the potential market. The goal is to accumulate market understanding as rapidly as possible. Hamel and Prahalad [38] recommend using expeditionary marketing as a way to learn fast, through a series of low-cost, fast-paced market introductions. The practical problem of this approach is how to reduce the time and costs of the iterations. The speed of the iteration is determined by the time it takes an organization to develop and launch a product or service, accumulate insights from the marketplace and recalibrate and launch again. Each iteration unfreezes one or more aspects of the product or service concept, and thereby provides an opportunity to apply what has been learned. The approach does not mean that an organization just launches products or services that are not ready or inappropriate to the needs of the users, but recognizes that user requirements in emerging markets can only be partially understood. Each iteration should embody everything that it is currently possible to know about user needs and desires.

It can be a challenge for organizations to adopt expeditionary marketing, as for many organizations it is not customary to celebrate 'failures'. The personal risk of failure is often too high for project managers or business managers to experiment. Managers seldom get punished for not trying, but they often get punished when they do try and do not achieve the target, even if they are relatively close. Another complicating factor is that in many organizations the manager only gets approval for investments if the business case is promising enough. This often leads to the situation that no approval is given if the projected sales quantities are not high enough. But for the expeditionary process the quantities should be kept low on purpose to speed up the iterations. One way to overcome these hurdles is to approach radical innovations as ventures. Although not a single organization will

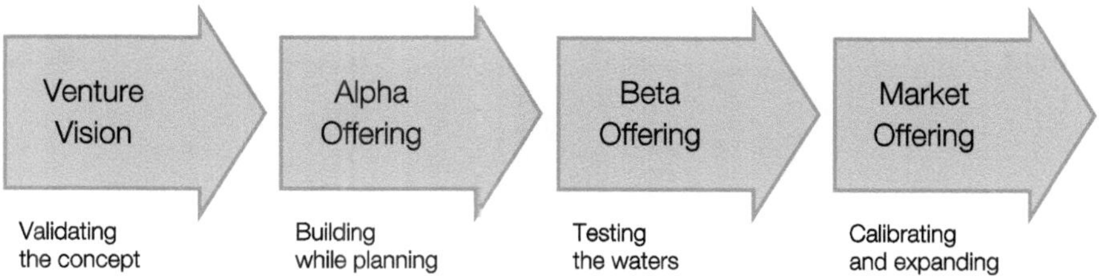

Fig. 5.13 The Bell–Mason venturing process

be introducing a transformational innovation to the market completely on its own, the challenges that the members of the ecosystem have to convince their internal stakeholders (such as their managers), should be taken into account.

Venturing Process Ventures are early stage, high-risk, high-potential, start-up companies. They follow a different process from the traditional milestone-driven processes to create new business in established companies. Private venture capitalists invest in these new businesses to capture part of the value once the business becomes a success. They accept that a number of businesses will fail, but the rewards from the successful businesses are high enough to make it a profitable activity. They have an entrepreneurial spirit, and often invest in businesses of which they have no knowledge themselves. This gives the managers of the ventures in which they invest more freedom of movement. This is important for meaningful innovations, as it presents ways of working that will increase their likelihood of success.

The Bell–Mason Venture Development Framework has proved to be a practical model to support organizations in corporate venturing. The framework provides an objective means to evaluate the progress of early-stage ventures, built on a large base of 'best practices' [52]. Large companies like Philips have adopted this approach to evaluate the progress of their internal incubator and new business projects, as an alternative to the milestone-driven processes applied for the incremental innovation projects in their existing businesses. By adopting a different process, these projects are given a fair chance to survive the initial phases and to grow and prove their potential for the future.

The framework divides the process into four distinct stages: venture vision, alpha offering, beta offering and market offering, as shown in Fig. 5.13. The process starts when a concept is available, while in the final stage the business is being built. It is very suitable for radical innovations, as it is used to handle complex, unstructured, experimental and always shifting ventures [52].

To quantify the progress of a venturing project, incremental performance measures are defined for each stage covering twelve dimensions: platform, product/service, delivery, business plan, marketing, business development/sales, CEO, team, board, cash, financing and control/operations. All actions in the project relate to these twelve dimensions, and as the project develops the progress of each of the dimensions is monitored to check its status in relation to the ideal best

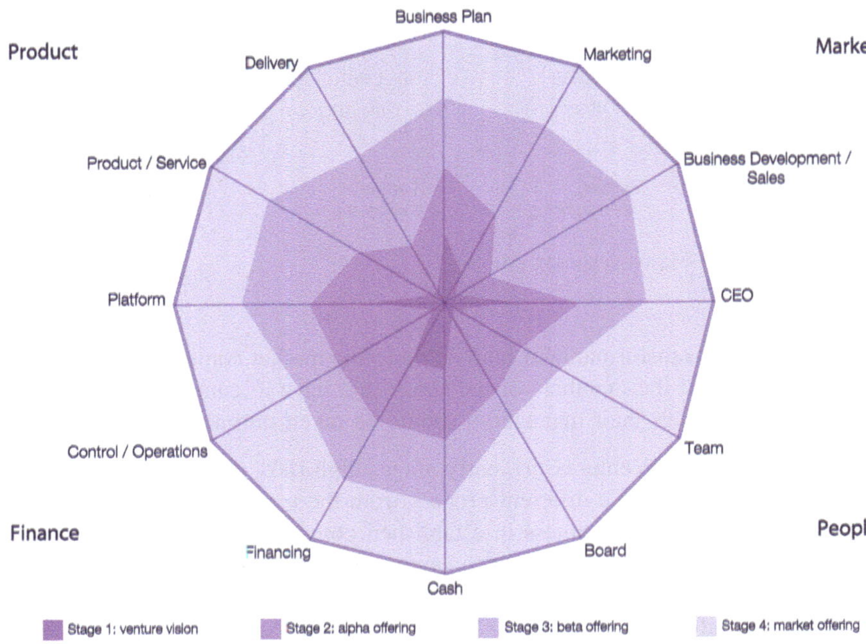

Fig. 5.14 The Bell–Mason venturing diagnostic with ideal states per stage [52], see also www.bellmasongroup.com

practices. These ideal states, shown in Fig. 5.14, are based on the analysis of over 1,600 best practices in venturing projects. By plotting the actual progress of a project on the spider graph, a team is able to get a clear overview of the current project status compared with the ideal. This helps to identify activities to improve the chances of success of the project.

The approach is interesting as it combines the assessment of hard facts with a human-focused assessment of the team and the working approach. The hard facts and figures are needed to assess the business potential and to get cash and financing from investors. But it only tells part of the story. Experience shows that a very critical success factor for new business is a team that has the drive to succeed despite all the problems that are encountered along the way. What is interesting is that the initial 'cowboy-like' behavior is needed to get through the initial stages of the project, but as the project progresses it becomes important to also prepare a sound business process. So in the assessment of the team, it is important also to see how this transition is made, and which additional people are coming on board to install the robust processes for the operation needed as business expands. The chemistry in the team and the diversity of characters are at least as important as the figures in the spreadsheets.

The framework has proved its success in new business creation, and is also used in breakthrough innovation projects in existing businesses, such as in Philips Consumer Lifestyle.

Although the twelve elements of the spider graph do not mention building an ecosystem, the model has proved its value in projects in which different parties work together to create a new business.

The venture process model helps to find a way to implement the value proposition and to develop it further into a market offering that is launched on the market. An important aspect that is not specifically addressed in the venturing approach is the design of new ecosystems. This topic will be discussed in the next chapter.

Designing Business Models for Transformational Innovations

A new business model can play an important role in bringing new products to the market. A business model can be specifically designed to take away initial hesitation and resistance to change among customers that manifest themselves as an unwillingness to adopt the innovation. One example of this was at the onset of the mobile phone market, when companies providing mobile telephone services offered their customers the device themselves ‘for free’. This made adoption easier for new customers, as they did not have to pay an initially high price for the phone itself. Especially if users are not yet familiar with a product or service, it can be helpful to lower the initial price. Of course the phones were not really ‘free’ at all, but the business model was built on a higher payment per call which was included in the subscriptions, so the initial costs would be covered by the recurring subscription payments.

Apple provides another example of business model innovation. Some people may say that products like the iconic iPod, iPhone and iPad are what have made Apple so successful in recent years. This is understandable, as these products are very visible and their design appeals to many people. The real innovation that started the success of the iPod was the digital rights management system that addressed the concerns of the music industry that they would lose control of their intellectual property. The digital rights management system built into iTunes allowed Apple to assure the industry that it would keep control, and at the same time Apple created a legal music download service that met the wishes of its customers. They got a way to browse hundreds of thousands of songs, and buy individual tracks or whole albums at very reasonable prices. So the real breakthrough was not so much the good product design of the iPod (although that certainly helped to make it a success), but the revolutionary business model: the iTunes music store. Apple subsequently developed iTunes to into an AppStore for use with the iPhone and the iPad, using the same business model. And more recently it created an AppStore for use with the normal Mac computers. The store lets people download legal apps of software packages or games at lower prices than buying the full software programs on DVD as official packages.

Business model innovation is about creating fundamentally new kinds of businesses, or about bringing more strategic variety into the business you are already in—the kind of variety that is highly valued by customers [67]. The example of Apple iTunes and iPod shows the power of combining hardware, software and services into a business model [40]. At first sight there does not seem to be a big difference between an Apple iPod and any other brand of MP3 player. But connecting it to iTunes adds not just software content and an online store to the system, but also the service of automatically updating firmware versions.

There is fast-growing literature on business model innovation, with more and more practical examples of how business models can be innovated. Most of the literature on business models focuses on single companies and their value propositions, but there is a growing notion that business models should also be considered in the context of networks and joint innovation. This is particularly of interest for the innovations that address societal needs, as these usually require a set of interdependent activities executed by a network of parties acting together. In such a system value is created in concert with other parties. This section starts by defining business models for single organizations, and later on also describes how business models for networked value propositions can be designed.

Why New Business Models Are Needed The characteristics of transformational innovations make it likely that new business models will be needed to turn their value propositions into a success. The literature gives several reasons for business model innovation, each linked one way or another to transformational innovations:

- Business models are perishable: many companies fail, not because what they are doing is wrong or mediocre, but because they keep doing what used to be the right thing to do for too long, and they fall victim to the rigidity of their own business models [24]. Success today does not guarantee success tomorrow [32]. This can either be because new categories of products or services are emerging (e.g. netbooks), or because social norms are changing (e.g. about sustainability), or because there are changes in the financial constraints (e.g. the price of gasoline). Such changes can render certain models more attractive and others unattractive [34]. Companies easily fall into the trap of 'vicious cycles': when revenues are slowing down they reduce headcount and expenses, which triggers further business declines, resulting in even more cuts. These 'vicious cycles' are hard to break; a better way out is to try new ideas and new business models. It takes courage to this, especially at times of financial difficulty, but it is the only way out [11]. So in the case of transformational innovations this means two things: on the one hand transformational innovations can be a way to get out of a vicious cycle and into a new virtuous one; on the other hand companies should be aware that the business models for transformational innovations are themselves perishable.
- Targeting growth: as Chesbrough remarked, a mediocre technology with a great business model may be more valuable than a great technology with a mediocre business model [10]. Much of the literature on business models is written from the perspective of one company, and with a focus on monetary value: profit. This can be illustrated by the definition of Teece: the essence of a business model is to deliver value to customers; enticing them to pay for this value; and

converting those payments into profit [72]. The profit formula is an essential part of the business model, defining how the company creates value for itself while providing value for its customers. Business models can create opportunities for growth by bringing an existing technology to a whole new market [40]. Alternatively companies can aim to jump from the maturity stage of one business to the growth stage of the next, thus creating new room to grow [58]. A business model generates a profit when activities are developed and resources are accumulated that drives a wedge between operating costs and revenues [28]. Even in existing markets revenues can be raised, when a more convenient way for customers to solve their problems or achieve their aspirations is found [39]. Open business models enable an organization to be more effective in creating value as well as in capturing it. They help to create value by leveraging many more ideas because of their inclusion of a variety of external concepts. They also allow greater value capture by utilizing a firm's key asset, resource or position not only in that organization's own operations but also in other companies' businesses [11]. For transformational innovations these more open business models are applicable, and are a way to target growth for the organizations involved in bringing the value proposition to the market.

- Emerging markets: one of the strategic conditions that require a business model change is when addressing the needs of large groups of potential customers, who are shut out of a market entirely because existing solutions are either too expensive or too complicated for them. These situations require a disruptive innovation, with a new business model [40]. When a new value proposition aims to create a new market, it needs to appeal to people who are currently not consuming, and to overcome the barriers that were preventing these people from consuming existing solutions. These barriers can be wealth, skills, access and time [39]. Emerging economies require different business models. The traditional business models are built on value propositions of products and services that require people to have a relatively high income, and in emerging markets the higher income tiers are simply not big enough to generate sufficient returns [25]. It makes more sense to target the vast middle market in emerging economies. The needs of those people are poorly addressed by the existing low-end solutions, and they cannot afford the cheapest of the high-end solutions. A value proposition for this middle market needs to combine affordability (and not selling less for less) with access (ability to reach dispersed populations). An example of a product that does just that is ChotuKool: a small portable fridge running on a cooling chip and a fan for about $69, in which small amounts of food can be kept for 2 days, even during power shortages [25]. Such solutions offer real business perspectives and are not just philanthropy. Nevertheless businesses entering emerging markets need to anticipate having to deal with imperfect markets, uncertain prices or costs, unreliable infrastructures, imperfect or absent governance, untested applications of technology and/or unpredictable competitive responses [73]. But visionary businesses can play a role in creating new business models that open up these new markets, and simultaneously attend to improving societal wealth. Collaboration between corporate

enterprises and non-governmental organizations often enables value to be created in novel ways, and minimizes costs and risks because such organizations bring in complementary capabilities not only in the form of tangible assets (human capacity, production capacity and market access), but certainly also in the form intangible assets (knowledge, reputation and know-how) [15].
- New business development: business models can also support new ventures in allowing entrepreneurs to explore a market and in constructing networks of technology providers and economic parties. In this context business models have a more prospective role in envisioning the future venture and the value creation logic that it will entail. The business model helps to articulate the value proposition and to find partners. It is used in reflective exercises to enhance the rationality of the proposition. For this purpose the business model should enable the process of collective exploration, be flexible enough to prevent lock-in and allow for the inclusion of alternative scenarios and actors [21]. A business model may help entrepreneurs or start-ups to 'sell' their business to stakeholders, or to resource holders who can become stakeholders when they recognize the value of the new venture [79]. When a new value proposition aims to create an industry discontinuity, it usually also requires a new infrastructure, such as for the introduction of electric cars [39]. In such cases more stakeholders will certainly need to adopt the innovation to realize the required infrastructure and complementary products and services.

Transformational innovations have characteristics that are a combination of the challenges described above. They are like new ventures in the sense that they aim to bring novel solutions. And they are like businesses for emerging markets in that they deal with imperfect conditions and target customers for whom there has not yet been any suitable solutions. At the same time they aim for a sound business as well, and for the organizations involved they offer a way to find new growth opportunities that may potentially disrupt their existing businesses. Designing business models for transformational innovations will therefore require an exploratory approach.

Transformational innovations are also very likely to be developed in a value network or value constellation. In this notion, all the actors in the value constellations need to be embraced, and the business model is used to assess the interest of actors to participate in a particular configuration. For this purpose the business model is told as a story to the stakeholders from multiple enterprises. It is important that the different stakeholders interpret the model in the same way [31].The definition of business models, which up to now has largely been targeted at single firms, will need to be extended toward whole ecosystems.

Transformational innovations address not only monetary value. As well as profit, other values need to be explicitly addressed to be suited for the purpose of meaningful innovations. It is very likely that the ecosystem will consist of a combination of organizations that strive for profit and non-profit organizations. The innovation will need to provide value, tangible and/or intangible, to each member of the ecosystem. The 'numbers test' to see if the business model is sustainable will also need to address these intangible values. But the idea is the same: the incurred costs (including intangible assets) will have to be recovered by value flowing back to the parties that invested, with at least slightly more value flowing in than was invested, to ensure sustainability of the business model.

The process of creating value through transformational innovations starts with an initial value proposition. Business modeling can then be a great help in understanding the value of the total value proposition (without already looking into the different products and services that are combined into the total solution). This also means that it does not (yet) matter which of the ecosystem members offer which part of the solution. For this first step the solution is considered as a single entity, and so is the consortium of organizations providing the solution. The purpose of business models at this stage is to check the value of the total solution for the customers and users, or in other words to allow the entrepreneurs to explore the market and how value can be created. To put it differently: the aim at this stage is to make the whole cake as big as possible. Icing and cherries can be added later, and after that the discussion on dividing it into pieces can start. But it is important first to get the core of the proposition right. Business modeling can be a big help at this stage.

Detailing the Value Proposition The first step in designing a business model is to further detail the value proposition. The initial value proposition describes for whom value is created, what the needs or desires of this target group are, what is offered as a solution, what alternatives are available and why this solution is more compelling than the alternatives. Such a value proposition does not describe how the value is actually delivered to the target group. At a more detailed level there are a number of choices to be made that will further clarify how the business will eventually work out. Different options can be selected on several design parameters [2, 3, 37, 39, 56]. These design parameters can be grouped into three categories: parameters related to the offering itself; parameters related to the delivery of the offering and parameters related to the payment.

Before going into details of the business model, it is important to understand that value propositions may be altered when considering the business model and the alternative solutions. Often the exercise of detailing the value proposition highlights issues that require reconsideration of choices made earlier. This is an inherent consequence of the iterative and explorative nature of designing new business models for new value propositions. An example is the initial value proposition to address the millennium goal of reduced mortality of pregnant women and their babies in rural areas of developing countries. This was to create a device that supports community health workers by providing information on the importance of hygiene, nutrition and maternity-related health issues. The focus of the initial value proposition was on the device, which was wirelessly connected to get up-to-date information, with the ability to present it to a small group of pregnant women at the same time. Figure 5.15 shows a sketch of the initial value proposition in the Savera project as initiated by Design Initiatief [20].

When considering the business model for such a device, it became clear that the original value proposition would not be feasible as neither the patients nor the

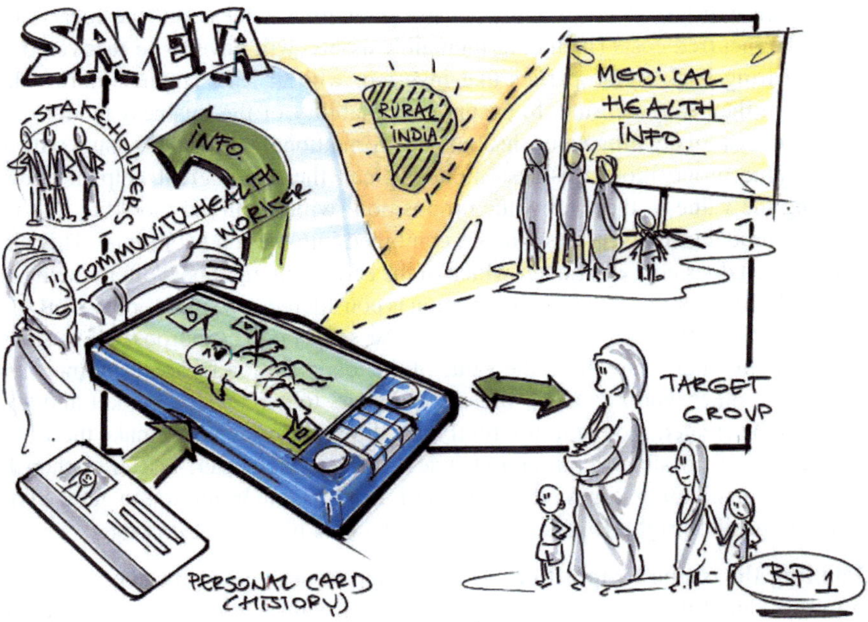

Fig. 5.15 Initial value proposition of the Savera project [20]

community workers would be able to spend the money to buy the device. At the same time it became clear that the initial value proposition could be enriched by adding information-based services. An important service would be for the community health workers to discuss specific cases with hospital doctors, who would be able to make better diagnoses of specific issues and provide personal advice for the patients through the health workers. Another service would be to make timely appointments with the hospital doctors when patients required them. This would greatly enhance the service to patients, who have to travel relatively great distances, and without appointments would have to wait in the streets around the hospital until they could be seen, with the obvious risks to their health. Adding these services to the value proposition made other revenue models possible, as now doctors and hospitals, and with them insurance companies, came into the picture as potential parties that might want to spend money on the proposition. This example also indicates that the process of designing the business model for meaningful innovations automatically leads to a creative process to enhance the value proposition. This creative process can lead to additional value for additional stakeholders, and therefore to a bigger overall cake to be shared among them.

Design Parameters Related to the Offering There are several options to choose when considering the offering in the value proposition. Is the offering a single product or service, or a bundle of several products and services? Will customers be able to choose from a range of products, or will there be a standard offering for all

Type	Product	Service
Features	Focused *Limited set of features*	Extensive *Extensive set of features*
Form	Tangible *Measurable or concrete product or service attributes*	Intangible *Intangible attributes*
Range	Narrow *Single product or service*	Broad *Full range of products and services*
Customization	General *No ability to be customised*	Custom *Can easily be customised for individual needs or preferences*
Breadth	Single *Single product or service offered to customer*	Bundled *Multiple products/services delivered to customer simultaneously*
Lifetime	Consumable *Consumed multiple times / used once*	Durable *Consumed once / used multiple times*
Target	Consumer *Consumer products or services (B2C)*	Business *Professional market(B2B)*
Quality	Cost driven *Focus on costs*	Quality driven *Focus on value and quality*

Fig. 5.16 Design parameters related to the offering

customers, or will there even be an option to customize the offering? Will the focus be on costs of the offering, or is it aimed more at the high-end segment, focusing on value and quality? Is the offering presented as a consumable, with frequent repeat purchases, or is it a durable that will be used multiple times? Figure 5.16 gives an overview of the design parameters related to the offering.

The list of design parameters can be used to check whether these are clearly defined, but they can also be used to inspire a new round of ideas that could enrich the value proposition. A discussion on the breadth of the value proposition can make ideas pop up for a bundled proposition, where initially the idea might have been a single product or service. It is important to keep in touch constantly with the insights that were gathered on the users and customers, and not to lose track of what provides true value for them. It is easy to get lost in making the proposition bigger. And although it might seem that offering more means more value, the complexity could create additional risks in realizing the proposition. A more complex system of products and services can be bigger, but does not necessarily provide more value.

Design Parameters Related to the Delivery of the Offering The second set of design parameters relate to how the offering is delivered to the target customers, as shown in Fig. 5.17. These parameters also define how accessible the offering will be to the customers.

Distribution	Direct *The offering is supplied directly to the customer*	Indirect *The offering is supplied by a third party*
Support	Light *The customer is provided with little or no support while purchasing*	Heavy *The customer is provided significant purchasing support*
Frequency	Often *The offering is purchased frequently*	Rarely *The offering is purchased infrequently*
Availability	On-demand *The offering is available all the time*	Limited *The offering is available at limited times*
Timing	Immediate *The offering's value is realised immediately*	Delayed *The offering's value is not realised immediately at the time of purchase*
Autonomy	Assisted *The offering requires assistance to use*	Independent *The product is used by the customer without help*
Strategy	Planned *Purchasing the offering is pre-mediated*	Unplanned *Purchasing the offering is spontaneous*
Location	Traditional *The offering is purchased in the same location as old solutions*	New *The offering is purchased in a new location*
Channels	Single *The offering is offered through a single channel*	Multi *The offering is offered through multiple channels*
Coverage	Local *The offering is only available locally*	Global *The offering is available worldwide*
Breadth	General *The offering is targeted at the general market*	Selected *The offering is targeted at a niche market or at a specific market segment*
Relation	Transactional *The exchange of the offering is purely transactional*	Relational *The offering is based on a strong relation with the customer*

Fig. 5.17 Design parameters related to the delivery of the offering

Making choices in these parameters triggers discussion of aspects like the channels through which the product or service will be available—will these be traditional channels such as regular shops, or will they be webshops. Of course the choices made in these parameters should be consistent with the overall strategy and aim of the value proposition. When offering a high-end product, it makes more sense to sell it in a store where the target customers would expect to find it, and where the atmosphere in the store matches the high-end level of the offering. Or when offering highly customizable and more complex value propositions, it makes sense to offer support during the purchasing process, as well as the product itself. This can be face-to-face by trained staff in a shop, but it could also be done online through a smart wizard, such as the online configurators that are commonly provided for cars.

The design parameters also address the kind of market that is targeted. Will the product be available to the general market worldwide, or will it be offered to a specific target group in a specific region. The product might also initially be aimed at a specific group, with a roadmap for later extension of the channels through which the product or service is offered. This could be a way to keep the risks under control. Feedback from a specifically selected small target group could then be used to improve the offering, and to increase its availability to larger groups of customers only when the uncertainty has been reduced.

A coherent selection of options makes an offer more compelling and accessible for customers. Although the decisions on these parameters are often made from the perspective of the organization or the ecosystem that markets the offering, there should be a continuous consistency check of the selections made from the customer's point of view. What looks logical to the organization may not be so logical for the customer. Especially for really new value propositions and new business models, there can be big uncertainties about what prospective customers expect, and where they expect to get such products and services. Exploring the options is also recommended for these decisions. When the value proposition is checked with prospective customers, the check on the design parameters for the delivery of the offering can be done at the same time.

Design Parameters Related to the Payment The third category of design parameters deals with the options related to payment for the value proposition. A customer pays a price for a product or service. Even if the product looks free of charge, there is still a hidden payment. In the case of intangible media such as online and mobile services this happens by presenting advertisements to the customer, who 'pays' for the service by viewing the advertisement. This is an indirect payment by the advertiser to the provider of the online services. Money is always an important aspect of any business model. Even in social business models, where social rather than economic profit is the aim, the right balance needs to be found between economic and social value [15].

Other parameters related to the payment include pricing, with a choice of either a standard, fixed price for the offering, or a variable price that is set for each individual transaction. The payment can be made either directly in cash or through a financing scheme, probably in installments. Figure 5.18 summarizes the various design parameters relating to payment.

By going through the design parameters, it becomes clearer what is being offered and how it is offered to the users. Discussing these aspects may also make it clear that certain aspects are still open, or that there is a lack of knowledge to help choose between the alternatives. In the service solution for maternity care referred to earlier, a discussion arose of who would pay for the service. Would the pregnant women themselves be willing (and able) to pay for the service, or would another payment scheme be more likely? As medical insurances are not yet commonplace in rural India, these would not be the paying customers. But in discussing the options it became clear that a form of 'micro-insurance' (in which people insure their health partly and for a limited time by making small payments)

Form	Exchange *The offering is purchased with cash*	Finance *The offering is purchased through financing*
Price	Fixed *The price of the offering is fixed*	Variable *The price of the offering can vary (e.g. auction)*
Unit	Pay per item *Customer pays for each item*	Pay for value *Customer pays proportionally for the value extracted*
Frequency	Once *The customer makes one complete payment*	Instalment *The customer makes more, smaller payments*
Timing	Up-front *Customer pays before receiving the offering*	Afterward *Customer pays after receiving the offering*
Payer	Single *One customer pays for the entire offering*	Multiple *Multiple customers pay for the entire offering*
Retention	Seduction *Customer is seduced to repeat purchase through the price & quality of the offering*	Lock-in *Lock-in model is used to enforce repeat purchase*
Medium	Tangible *Customer pays with a tangible medium such as money or goods*	Intangible *Customer pays with an intangible medium such as knowledge or attention*

Fig. 5.18 Design parameters related to the payment

was becoming more popular. Connecting the value proposition to these micro-insurances certainly makes the total value proposition more feasible.

Several scenarios can be explored in the design parameters. Once decisions are taken on what is to be offered and how it will be offered, the next step is to define a value formula for the business model.

Defining the Profit Formula In business model literature the terms 'profit formula' or 'revenue model' are usually used. The aim in this book is to design solutions for societal challenges, with innovations that aim to deliver more value than financial profit alone. The term 'value formula' is therefore used in this context. In the value formula, the revenue formula is extended beyond monetary value, as will be shown in the next paragraph. But monetary value is still a crucial aspect, also in meaningful innovations, so a good understanding of the revenue model is important. Plenty of different revenue models are described in the literature, see e.g. [34, 39, 49, 59, 64, 72, 75]. These give rather generic descriptions of the business models used by organizations. They can serve as exemplar models that can be copied or imitated [21], or as recipes that demonstrate and advise on how to do something [1]. A business model used by one company in a specific environment can be copied by another organization in another environment. At a superficial level a business model may often look easy to imitate, and the basic idea and business logic behind a new model are unlikely to enjoy intellectual property protection. Even descriptions of business models that might be protected by copyright are unlikely to be a barrier to copying the core idea. So in that sense copying a business model is not a problem. In practice, however, business models require systems, processes and assets that may be far more difficult to replicate. It is not always transparent to outsiders how the business model is implemented, or why customers accept a specific model [72]. But other business models can still be a great inspiration in designing the business model for a new value proposition. The choices made should be logical with respect to the target customer group and their abilities and accessibility. Different revenue models offer different opportunities, and it is very likely that revenue models can also shift over time. To attract new customers to a new value proposition you could choose a revenue model built on a 'freemium' or a 'reverse bait and hook' model. Combined with a 'lock-in' model this can make it harder for customers to switch to alternative solutions, and ensures a continuous stream of revenues that can be used to extend the value proposition and offer more 'bundled' products and services that increase the value of the total solution.

Rita Gunter McGrawth [35] advises looking for models that create customer 'stickiness', loyalty or barriers to entry. Anything that automatically gets renewed creates a certain amount of stickiness, especially if customers have to work to change providers. Another profitable model is the platform model, such as that used by Microsoft. Microsoft of course sells its software to end-users, but it also makes a lot of money by licensing its software platforms to other organizations. In this way it can charge more, and can hold on to an advantage for a long time.

There are also business models that should be avoided if possible: such as those in which customers buy something once only. Even for social innovations that aim for social value, it is important to make sure that the revenue model is clear; there should be a clear path to monetization.

At this stage, the main concern is to find a revenue model that makes the total solution attractive for the customers, and feasible at the total value network level. The value network then looks like just a single organization. The revenue model discussion then focuses on the revenue streams between the customers and this aggregate organization.

Extending toward a Value Formula The revenue model is the core of the business model, even for social innovations. The economics are important to achieve a sustainable situation. Even non-profit organizations cannot sustain their business if their costs are not covered. However, they might have less direct payments by users, with a business model built on subsidies from the government or charity organization. So money streams are a very important element of the value formula, but for meaningful innovations they are only a part of the whole story. As well as money, other intangible values also play an important role. These values are important to both parties involved in the exchange: the customer on the one hand and the offering organization on the other hand.

For the customer, the intangible value is included in the description of the value proposition, and is an integral part of the total solution that is offered. But the intangible value may be less clear for the organization providing the value proposition. As in the revenue model, it should be made clear how the intangible value flows toward the organization. Examples of intangible value are knowledge, reputation, exposure or attention. In the value formula, the flow of such value from the customer toward the organization should be made explicit. For example, if a customer who buys a solution in the healthcare domain discloses information on his lifestyle, eating habits or other relevant aspects, that information can be of great value to the organization. Similarly, when customers voice their positive experiences with a product or service to other users, this builds the reputation of the organization. For non-profit organizations, values of these types can sometimes be even more important than the financial revenues.

Value-sharing in Ecosystems When more organizations collaborate in a value network to bring a new value proposition to the market, it is also important that the value formulas of the separate organizations are sustainable. The revenues and values generated by the total solution will have to be divided and distributed among the various members of the ecosystem. This issue will be explained in more detail in the next chapter that deals with the design of new ecosystems.

References

1. Alblas, Alex, Elke den Ouden, and Rens Brankaert. 2011. Creating reciprocal value propositions. The case of the improvement of the quality of life of Alzheimer patients and their caregivers. Submitted for the 4th world conference on design research, International Association of Societies of Design Research, 31 October–4 November, Delft, The Netherlands.
2. Baden-Fuller, Charles, and Mary S. Morgan. 2010. Business models as models. *Long Range Planning* 43: 156–171.
3. Ballon, Pieter. 2007. Business modelling revisited: The configuration of control and value. *Info* 9(5): 6–19.
4. Behrens, Roy R. 1998. Art, design and gestalt theory. *Leonardo* 31(4): 299–303.
5. Bielderman, Marlies, Mili Docampo Rama, Megumi Fujikawa and Lekshmy Parameswaran. 2008. Envisioning transitions: A design process for healthcare and wellbeing. Philips Design.
6. Blamey, R.K., Mick S. Common, and J. Quiggin. 1995. Respondents to contingent valuation surveys: Consumers or citizens? *Australian Journal of Agricultural Economics* 39(3): 263–288.
7. Blomquist, Asa, and Mattias Arvola. 2002. Personas in action: Ethnography in an interaction design team. Proceedings of the NordiCHI, 19–23 October.
8. Brown, Tim. 2009. *Change by design: How design thinking transforms organisations and inspires innovation.* New York: HarperCollins Publishers.
9. Button, Graham. 2000. The ethnographic tradition and design. *Design Studies* 21: 319–332.
10. Chesbrough, H. 2010. Business model innovation: Opportunities and barriers. *Longe Range Planning* 43: 354–363.
11. Chesbrough, Henry W. 2007. Why companies should have open business models. *MIT Sloan Management Review* 48(2): 22–28.
12. Citizen M. Philips 2007. Introduces simplicity to the hotel experience. Press release, 23 Oct. 2007 http://www.citizenm.com/press/philips-introduces-simplicity-to-the-hotel-experience.
13. Cooper, Alan. 2004. *The inmates are running the asylum: Why high tech products drive us crazy and how to restore the sanity.* Indiana: Sams Publishing.
14. Cross, Nigel. 2006. *Designerly ways of knowing.* London: Springer.

15. Dahan, Nicolas M., Jonathan P. Doh, Jennifer Oetzel, and Michael Yaziji. 2010. Corporate-NGO collaboration: Co-creating new business models for developing markets. *Long Range Planning* 43: 326–342.
16. Ulrike, De Brentani. 2001. Innovative versus incremental new business services: Different keys for achieving success. *Journal of Product Innovation Management* 18: 169–187.
17. Marieke, De Mooij. 2004. *Consumer behavior and culture: consequences for global marketing and advertising*. London: Sage Publications.
18. Den Ouden, Elke. 2009. Designing (Added) Value: Social Innovation in Flexible Networks. Inaugural Lecture Eindhoven University of Technology. ISBN 978-90-386-2096-1, 23 October.
19. Den Ouden, Elke, Darren Ee, and Nicky Goh. 2008. The Philips InnoHub: Generating Breakthrough Innovations in an Open Innovation Setting. PDMA Visions Magazine, March.
20. Den Ouden, Elke, and Rianne Valkenburg. 2011. Balancing Value in Networked Social Innovation. Participatory Innovation Conference Proceedings, Sonderborg, Denmark, 13–15 January.
21. Doganova, Liliana, and Marie Eyquem-Renault. 2009. What do business models do? Innovation devices in technology entrepreneurship. *Research Policy* 38: 1559–1570.
22. Dorst, Kees. 2003. The problem of design problems. In *Expertise in design*, eds. N. Cross, and E. Edmonds, 135–147. Sydney: Creativity and Cognition Studio Press.
23. Dourish, Paul. 2006. Implications for design. CHI 2006 Proceedings Design: Creative & Historical Perspectives, Montreal, Quebec, Canada, April 22–27.
24. Doz, Yves L., and Mikko Kosonen. 2010. "Embedding strategic agility: A leadership agenda for accelerating business model renewal". *Long Range Planning* 43: 370–382.
25. Eyring, Matthew J, Mark W. Johnson, and Hari Nair. 2011. New Business Models in Emerging Markets. Harvard Business Review, January.
26. Fischer, Gerhard, and Elisa Giaccardi. 2006. Meta-design: A framework for the future of end-user development. In *End user development. Human computer interaction series*, vol. 9, eds. Henry Lieberman, Fabio Paterno, and Volker Wulf, 427–457. Berlin: Springer.
27. Gerd, Fricke. 1999. Successful approaches in dealing with differently precise design problems. *Design Studies* 20(5): 417–429.
28. Gambardella, Alfonso, and Anita M. McGahan. 2010. Business model innovation: General purpose technologies and their implications for industry structure. *Long Range Planning* 43: 262–271.
29. Garcia, Rosanna, and Roger Calantone. 2002. A critical look at technological innovation typology and innovativeness terminology: A literature review. *Journal of Product Innovation Management* 19: 110–132.
30. Goel, Vinod, and Peter Pirolli. 1992. The structure of design problem spaces. *Cognitive Science* 16: 395–429.
31. Gordijn, Jaap, Alexander Osterwalder, and Yves Pigneur. 2005. Comparing two business model ontologies for designing e-Business models and value constellations. 18th Bled eConference eIntegration in Action, Bled, Slovenia, June 6–8.
32. Govindarajan, Vijay, and Chris Trimble. 2011. The CEO's role in business model reinvention. Harvard Business Review, January/Febrauary.
33. Griffin, Abbie. 2005. Obtaining customer needs for product development. In *The PDMA handbook of new product development*, 2nd edn, ed. Kenneth B. Kahn. New York: Wiley.
34. Gunther McGrath, Rita. 2010. Business models: A discovery driven approach. *Long Range Planning* 43: 247–261.
35. Gunther McGrawth, Rita. 2011. When your business model is in trouble: How to recognise—and react to—signs of an impeding crisis. Harvard Business Review, January.
36. Gunther McGrath, Rita, and Ian C. MacMillan. 1995. Discovery driven planning. Harvard Business Review, 1995.
37. Haaker, Timber, Edward Faber, and Harry Bouwman. 2006. Balancing customer and network values in business models for mobile services. *International Journal of Mobile Communications* 4(6): 645–661.

38. Hamel, Gary, and C.K. Prahalad. 1994. *Competing for the future: Breakthrough strategies for seizing control of your industry and creating the markets of tomorrow*. Boston, MA: Harvard Business Press.
39. Johnson, Mark W. 2010. *Seizing the white space: Business model innovation for growth and renewal*. Boston, MA: Harvard Business Press.
40. Johnson, Mark W., Clayton M. Christensen, and Henning Kagermann. 2008. Reinventing Your Business Model. Harvard Business Review, December.
41. Kelley, Tom. 2002. *The art of innovation: Lessons in creativity from IDEO, america's leading design firm*. London: HarperCollins Business.
42. Kim, W., and Renee Mauborgne Chan. 2005. *Blue ocean strategy: How to create uncontested market space and make the competition irrelevant*. Boston, MA: Harvard Business Press.
43. Kyffin, Steven, and Paul Gardien. 2009. Navigating the innovation matrix: An approach to design-led innovation. *International Journal of Design* 3(1): 57–69.
44. Louridas, Panagiotis. 1999. Design as bricolage: anthropology meets design thinking. *Design Studies* 20: 517–535.
45. Lawson, Bryan, and Kees Dorst. 2009. *Design expertise*. Oxford: Elsevier.
46. Lindsay, Christina. 2003. Involving people as co-creators. In *The new everday: Views on ambient intelligence*, ed. Emile Aarts, and Stefano Marzano. Rotterdam, The Netherlands: 010 Publishers.
47. Lynn, Gary S., Joseph G. Morone, and Albert S. Paulson. 1996. Marketing and discontinuous innovation: The probe and learn process. *California Management Review* 38(3): 3–37.
48. Maarse, Johan Henk. 2011. Cradle to cradle implementation at desso. Participatory Innovation Conference Sonderborg, Denmark, 13–15 January.
49. Malone, Thomas W., Peter Weill, Richard K. Lai, Victoria T. D'Urso, George Herman, Thomas G. Apel, and Stephanie L. Woerner. 2006. Do some business models perform better than others? MIT Sloan School of Management Working Paper No. 4615–06.
50. Martin, Roger. 2009. *The design of business: Why design thinking is the next competitive advantage*. Boston, MA: Harvard Business School Press.
51. Martin, Roger. 2007. *The opposable mind. How successful leaders win through integrative thinking*. Boston, MA: Harvard Business Press.
52. Mason, Heidi, and Tim Rohner. 2002. The venture imperative: A new model for corporate innovation. Harvard Business Press, Boston, MA. www.bellmasongroup.com.
53. Max-Neef, Manfred A. 2005. Foundations of transdisciplinarity. *Ecological Economics* 53: 5–16.
54. McDermott, Christopher M., and Gina Colarelli O'Connor. 2002. Managing radical innovation: An overview of emergent strategy issues. *Journal of Product Innovation Management* 19: 424–438.
55. Moore, Geoff. 2004. The fair trade movement: Parameters, issues and future research. *Journal of Business Ethics* 53: 73–86.
56. Morris, Michael, Minet Schindehutte, and Jeffrey Allen. 2005. The entrepreneur's business model: Toward a unified perspective. *Journal of Business Research* 58: 726–735.
57. Mullins, John, and Randy Komisar. 2009. *Getting to plan B. Breaking through to a better business model*. Boston, MA: Harvard Business Press.
58. Nunes, Paul, and Tim Breene. 2011. Reinvent your business before It's too late. Harvard Business Review, January.
59. Osterwalder, Alexander, and Yves Pigneur. 2010. *Business model generation: A handbook for visionairies, game changers, and challengers*. New York: Wiley.
60. Oxford Dictionary of English (revised edition). ed. Catherine Soanes and Angus Stevenson. Oxford University Press, 2005. Oxford Reference Online. Accessed 7 December 2010.
61. Parameswaran, Lekshmy, and Jeroen Raijmakers. 2010. People-focused innovation in healthcare. How philips design supports development of solutions for the ever-changing healthcare landscape. Philips Design.
62. Prahalad, C.K., and Venkat Ramaswamy. 2004. *The future of competition: Co-creating unique value with customers*. Boston, MA: Harvard Business Press.

63. Pruitt, John, and Jonathan Grudin 2003. Personas: practice and theory. Proceedings of the Conference of Designing for User Experiences, New York.
64. Rappa, M.A. 2004. The utility business model and the future of computing services. *IBM Systems Journal*, 43(1): 32–42
65. Rogers, Everett M. 2003. *Diffusion of innovations*, 5th edn. New York: Free Press.
66. Rotte, Andre, Rianne Valkenburg, Slava Kozlo, Otto Kokke. Future Telling. (In Dutch). Design Initiatief, The Netherlands, 2009.
67. Skarzynski, Peter, and Rowan Gibson. 2008. *Innovation to the core: A blueprint for transforming the way your company innovates*. Boston, MA: Harvard Business Press.
68. Smets, Gerda. Vormleer. De Paradox van de Vorm. (In Dutch) Uitgeverij Bert Bakker, Amsterdam, The Netherlands, 1986.
69. Smith, Preston G., and Donald G. Reinertsen. 1998. *Developing products in half the time: New rules, new tools*, 2nd edn. New York: Wiley.
70. Smith, Robert P., and Primanata Tjandra. 1998. Experimental observation of iteration in engineering design. *Research in Engineering Design* 10: 107–117.
71. Song, X.Michael, and Mitzi M. Montoya-Weiss. 1998. Critical development activities for really new versus incremental products. *Journal of Product Innovation Management* 15(2): 124–135.
72. Teece, David J. 2010. Business models business strategy and innovation. *Long Range Planning* 43: 172–194.
73. Thompson, James D., and Ian C. MacMillan. 2010. Business models: Creating new markets and societal wealth. *Long Range Planning* 43(2–3): 291–307.
74. Thompson Klein, Julie. 2004. Prospects for transdisciplinarity. *Futures* 36: 515–526.
75. Timmers, Paul. 1998. Business models for electronic markets. *Electronic Markets*, 8(2):3–8
76. Ulrich, Karl T., and Steven D. Eppinger. 2000. *Product design and development*. Boston: McGraw-Hill.
77. Valkenburg, Rianne, and Elke Den Ouden. 2011. Design Initiatief first phase. Working paper.
78. Valkenburg, Rianne, and Kees Dorst. 1998. The reflective practice of design teams. *Design Studies* 19(3): 249–272.
79. Verstraete, Thierry, and Estele Jouison. 2006. Connecting stakeholders theory and conventions theory to highlight the adhesion of stakeholders to the business model of a start-Up. *International Council for Small Business, 51st World Conference*, Melbourne, Australia, June.
80. Veryzer Jr, Robert W. 1998. Discontinuous innovation and the new product development process. *Journal of Product Innovation Management* 15: 304–321.
81. Hippel Eric, Von. 2006. *Democratizing innovation*. Cambridge, MA: MIT Press.
82. Wasson, Christina. 2000. Ethnography in the field of design. *Human Organisation* 59(4).
83. Wijnberg, Nachoem M. 2000. Normative stakeholder theory and aristotle: The link between ethics and politics. *Journal of Business Ethics* 25(4): 329–342.
84. Woodall, Tony. 2003. Conceptualising 'value for the customer': an attributional, structural and dispositional analysis. *Academy of Marketing Science Review* 2003(12): 1526–1794.
85. Yunus, M., B. Moingeon, and L. Lehmann-Ortega. 2010. Building social business models: Lessons from the grameen experience. *Long Range Planning* 43: 308–325.
86. Zaltman, Gerald. 2003. *How customers think. Essential insights into the mind of the market*. Boston, MA: Harvard Business Press.

Chapter 6
Designing Ecosystems

Introduction

The term ecosystem is increasingly being used to emphasize that innovation is no longer an activity that organizations can do in isolation. Many innovations are not just solitary products, but combinations of products and services in a larger system. In many cases different organizations are involved in these larger systems, as described in Chap. 2. Before considering the design of ecosystems, it is important to properly understand what ecosystems mean for transformational innovation.

Ecosystems for transformational innovation aim to bring together those organizations and stakeholders that affect or are affected by the innovation. The purpose of the ecosystem design is to ensure that the innovation will sustainably deliver the intended value to all members of the ecosystem. Three types of ecosystems are described in the literature: at the regional, industry and specific innovation project levels. The ecosystems for transformational innovations are comparable to those at the specific innovation project level.

These ecosystems are different from regional ecosystems, which focus on creating local clusters of different organizations that together create virtuous cycles and form rich pools of shared knowledge. Such clusters become magnets that bring together specialist knowledge from around the world [33], such as the concentration of information technology-related organizations in Silicon Valley [83]. Organizations can benefit from becoming members of such ecosystems by tapping into these innovation hotspots and taking advantage of the available talent [60] and specialized services [52]. These arrangements attract new entrepreneurs and enable the building of what Richard Florida has described as a 'global creative class' [36]: the more entrepreneurs gather in one place, the more attractive that place becomes to other like-minded people [60]. Some countries have a national strategy to become regional ecosystems, such as Finland with its investments in public education, flexible schemes for public and private players, and an innovation university to attract global talent and business for innovation and design. Another example is Biopolis in Singapore, where new businesses can participate in relationships with government

E. den Ouden, *Innovation Design*, DOI: 10.1007/978-1-4471-2268-5_6,

agencies, venture capital firms, global pharmaceutical companies, academic research labs and other institutions that share biotechnology-related resources [60]. In the context of this book, these regional ecosystems can be interesting places to tap into when developing new businesses for meaningful innovations [90]. However, the design of such ecosystem is beyond the scope of this book.

A second common notion of ecosystems is the industry ecosystem, which focuses on one specific industry. From the perspective of one organization and its strategic intent, a network of organizations is built around it to enable it to reach its long-term goals in a collaborative effort [60]. To some extent the inbound forms of open innovation can be seen as a way to set up such an ecosystem. These organizations set up informal and formal relationships to gain access to resources and the knowledge of partners [23]. The case of Boeing shows how ecosystems like this work. With the consent of customers, subcontractors, suppliers and other actors, Boeing is able to configure a more effective offering than the separate parties [86]. Sometimes companies even collaborate with their competitors in the innovation process [15]. A well-known example is when a new standard needs to be set up around new technologies (such as in consumer electronics for the CD, DVD and Blue-ray Disc formats). This works well in radical innovations where the still emerging markets mean there is less competitive pressure, and it is easier for the competing companies to differentiate their offerings because of their novelty. Collaborating competitive companies can benefit from their joint efforts to gain critical mass, and 'educate' the markets [88]. This type of ecosystem focuses on a specific industry and its business interest. However, its scope is too narrow for the purpose of meaningful innovations.

A third type of ecosystem is at project level, which is created around radically new products and services. Solutions for today's complex societal, economic and environmental problems transcend the ability of any one organization to solve them, and can only be addressed by collaboration between people and organizations from different sectors. The resulting innovations differ from other types of innovations in three ways: they are radical in nature by creating new markets while affecting the political, behavioral, social and technological environments; they may involve a high level of government influence because they offer far-reaching benefits for society; and they typically involve the convergence of old and new technologies [51]. In a joint effort, businesses, non-governmental organizations and public organizations are creating shared value by reconceiving products and markets. In these hybrid ecosystems, the concept of shared value blurs the line between for-profit and nonprofit organizations [83].

The project level ecosystem differs from the industry ecosystem in that it is built around a new value proposition (and not necessarily around an existing organization). Different scenarios are explored in designing such ecosystems, and the parties involved in the development of the value proposition and the value network that creates the ecosystem are not fixed members of the system. During the process, parties may decide to opt out because advancing insights make it clear that there is not sufficient value for them. New parties can also join because they bring relevant knowledge, experience or technology. If the project develops and becomes successful in the market, the ecosystem related to the project can evolve into a business level ecosystem.

This type of ecosystem is most relevant in the context of this book. When the term ecosystem or business ecosystem is used in the rest of the book, it refers to this project level ecosystem.

This chapter starts with a definition of ecosystems. As ecosystems are networks of a special kind, the relevant aspects of innovation in networks or value constellations as found in the literature will also be discussed. The second part of this chapter deals in detail with the design of new ecosystems, together with practical tools that can be used and examples from projects.

Defining Ecosystems

As discussed in Chap. 2, the term ecosystem stems from biology. There, ecosystems are described as both resilient structures that are able to adapt to changes in the environment, and as fragile structures that can collapse if changes occur [79].

There are a number of ideas in the concept of ecosystems that make them interesting as a metaphor for creating business through meaningful innovations. First of all, the notion that the members of the system are interdependent: they follow a process of co-evolution. Co-evolution in natural and social systems is a process in which interdependent species evolve in an endless reciprocal cycle—in which changes in species A set the stage for the natural selection of changes in species B, and vice versa. A second insight is that natural ecosystems sometimes collapse if environmental conditions change too radically. Dominant combinations of species may then lose their leadership as new ecosystems establish themselves. Plants and animals that formerly played a marginal role can then often become much more important [68].

As James Moore [69] observed, there are clear parallels and profound implications for current businesses dealing with the challenges of innovation. That makes the ecosystem model an attractive way to understand business. However, as in any model that aims at understanding dynamic response, there are limitations to their applicability. As the financial sector has learned, the model that was used in financial networks to understand the dynamics of bank networks paid insufficient attention to the possible effects on the stability of the system as a whole [50]. So although the metaphor is useful, care must be taken in its use for decision-making. In ecology species are single entities, but the typical members of business ecosystems have a much more complex structure and multiple relationships (such as organizations with many employees and departments, all having relationships at strategic, tactical and operational levels with other organizations).

In ecology, active ecosystem management is practiced to maintain viable populations of all native species. This requires management over periods of time long enough to maintain the evolutionary potential of species and ecosystems [45]. Similar management strategies can be adopted in business ecosystems to ensure the stability and sustainability of the system as a whole.

Ecosystems can be seen as a specific form of network or value constellation. Extensive literature is available on these subjects, and the following section gives a brief overview of the relevant aspects in this context.

Networks The notion of networks in the context of innovation has been gaining more attention since the rise of the knowledge economy. Networks offer a highly feasible way to use and enhance intangible assets such as tacit knowledge and technological innovation. Networks have an open-ended quality, which is most useful when resources are variable and the environment uncertain. If partners in a network are involved in ongoing, complementary activities, the relationship develops a level of trust and this is more likely to lead to the sharing of critical information. This in turn often leads to the emergence of common values [85]. This aspect of networks makes them important for innovations that require the combined knowledge of different parties.

Networks also provide a context for learning by doing. As information passes through networks it becomes both freer and richer: new connections and new meanings are generated, debated and evaluated by the complementary views of the partners in the network [85].

Networks flourish when there is a high level of trust. There is limited need for hierarchical supervision because the desire for continued participation successfully discourages opportunism. Monitoring is generally easier and more effective if it is done by peers than when it is done by superiors. Persons who are not willing to cooperate are punished by those who are. A person's reputation is an important asset in functioning in a network, and the most visible signal of their reliability. If trust declines, so does the willingness to enter into long-term collaboration. Cooperative attitudes are then replaced by calculative ones, and the need for formal agreements grows [85]. Trust is therefore a crucial aspect of successful networks. And although the networks for meaningful innovations will need to be built from scratch, it is important to keep the importance of trust building in mind.

From Company-Centered Networks to Solution Driven Networks Much literature uses a company-centered view of networks, connecting firms with outside co-producers [86]. For innovation, the network comprises the organizations with which the company collaborates to develop its new products. Typical questions are: With whom to collaborate?; Why collaborate?; and Are the skills available to make the collaboration work? [64].

Other authors emphasize that a company's network should be viewed as two external networks: those of its suppliers (the supply chain) and its buyers (the marketing, distribution and service providers network) [35]. In this industrial view, the value creation is seen as sequential, unidirectional and best described in 'value chains'. An alternative view to the chain is value co-production, where value creation is synchronous, interactive and involves co-creation by customers. The focus is then no longer on the company but on the co-produced offerings, which build on relationships with multiple actors [86]. If the focus shifts from the company itself to its offerings to the customer, it becomes clear that combining internal and external resources is a key strategic issue for survival, and a determining factor for successful business models [9]. A company-centered view forms an inadequate basis for understanding the dynamics and opportunities within such an interconnected world. A company-centered view also strengthens the illusion that the company controls the network, or more simply that it is its own network [47].

An example of an industry in which this shift is clearly visible is telecommunications. The established value chain is increasingly being deconstructed. Many tested business models have become obsolete due to rapid technological developments and market changes. The value chain is rapidly evolving into a value network with multiple entry and exit points, creating complexity because many organizations play all kinds of roles. Economics and value propositions differ significantly for different market positions and business models. For example banks are entering the telecom industry to provide mobile portals for their existing customers, with the aim of locking them in. Providers of wired services see mobile portals as a way to reach new customers. The roles of the different companies are continuously changing as new business models emerge [66].

If a market-driven approach to innovation is adopted, the network becomes the means to create more value by integrating offerings or competences from different organizations into better solutions for customers. Networks generate economic and social good by tangible and intangible value exchanges between workgroups, organizations or businesses [3]. Key innovators will have to set up and manage inter-organizational networks to develop and exploit business opportunities. External network management becomes crucial as a network of partners needs to be orchestrated and a unifying vision is instilled. Interesting partners need to be invited to the network, the network needs to be nurtured and tensions between partners need to be minimized [95]. These networks, sometimes also called business webs, are customer-centric and consist of legally independent but economically interdependent specialized organizations that contribute modules or knowledge to a product–service system. Business webs are often enabled by information and communication platforms [13].

When aiming for solutions that create systemic changes in society, there are several requirements for the network and its partners. The values in the network should be based on insights in the market, and should have a long-term focus to reach a common goal. The members should value integrity and tolerance, and balance top-down and bottom-up decisions, based on a deep understanding of the matter at hand. Although networks do not require hierarchy, they benefit from a connected leader with a long-term visionary approach. He or she achieves buy-in of the relevant stakeholders, thereby creating value for all and inspiring people to act according to the vision. Individual and collective interests should be aligned, and a 'win together' approach is used in strategic partnerships and in connection with society at large [96].

The strength of a value network is determined by the added value it provides compared with alternative solutions, and by the commitment of the participants [57]. Doing business in networks provides advantages because of the increased added value for the customers, the shared learning between the network partners, and the efficiency of being able to build on the specialized, complementary expertise and resources of the partners [65], but it is not without challenges.

Challenges of Working in Networks Managing innovation in networks is complex. Each of the organizations in a network attempts to manage its individual relationships, and to affect others elsewhere in the network with which they do not deal directly. Each organization has limited knowledge, and each has its own subjective view of the network, which drives its behavior [48]. Networks are built on tight and loose connections between the parties [86], differing in how interdependent the parties are. For successful collaboration, network firms must understand each other and each other's needs [65]. Nobody has the complete overview needed to understand in-depth how the network works and how it will react. Knowledge is incomplete, and the inadequacy of each company's knowledge means that learning by doing is an important aspect of networking [48]. There is no single, objective network. There is no 'correct' or complete description of it. The boundaries are not clear. The network is not owned by any of the parties, although all will try to manage it. No company is the hub of the network, and nor is there a center, although many organizations may believe that they have that role.

As Hakansson et al. [47, 48] describe: a company in a network is an actor whose identity is defined by others in the network and who is living a life within individual relationships and within numbers of connected relationships. This life is both interesting and complicated as it places the actor in a number of intricate paradoxes:

> **Opportunities and limitations** A network is both the sources of life for a company and the cage that imprisons it. The network as a resource constellation creates inertia that limits innovation, but it also creates a firm basis from which developments can take place. The existence of structure and its inertia makes action in the network more difficult, but also more important. The only way a company can achieve change is through the network. This requires persistence in convincing others of the benefits of that change and managing their expectations. Managers have to accept that change must often be accomplished within existing relationships, where some investments have already been made and where costs and benefits are more apparent. The question is more often how to interact with existing counterparts than about how to choose new ones. It is the network that provides the 'bundle' of different new and existing technologies that are necessary for any innovation. But the existing structure of the network acts as a brake on innovation because of the investment in existing ways of working and because of the requirement to enlist the cooperation of those with which the innovator does not have a relationship.
>
> **Influencing and being influenced** A company is both the determinant and the outcome of its relationships and what happens in them. A company should interact to try to understand how the network functions from the perspective of the specific others, and how it sees their positions and its own. No company has sufficient resources itself to satisfy the requirements of any customer. Interdependency between companies means that the strategy process is interactive, evolutionary and responsive, rather than independently developed and implemented. The party that is least committed to a relationship is likely to control it negatively by restricting its development. Conversely, the positive development of a relationship is likely to be driven by the party that is most committed to it. The co-determination of companies and relationships in a network means that the more important a company's relationships are, the more important will be the individuals who interact in

them. This emphasizes the importance for each company to manage all of its interactions carefully and for each individual person to interact self-consciously.

Controlling and being out of control The more a company achieves control, the less effective and innovative the network becomes. Companies should aim for control, but as soon as they acquire some 'final' control over the surrounding network they should be worried. Interaction in business networks leads to a process of learning and systematizing action. This takes advantage of the variety in the network and also capitalizes on the economy of network stability. This paradox reinforces the need for a manager to analyze his company's position in terms of its specific relationships and its own and others' resources, rather than in terms of a set of products, markets and competitors. It highlights the problems for managers if they take a self-centered view of the network. A network will look very different from the perspectives of different companies, each with its own motivations, resources and understanding. A company that only sees the network from its own perspective will fail to understand its dynamics and the interface between the well-being of others and itself.

These paradoxes give a clear view on the challenges that are related to doing business in a network setting

Innovation in Ecosystems

This section gives a more detailed description of ecosystems with a focus on those related to transformational innovations. It provides a background on why ecosystems are needed for these innovations, and how they work.

Creating Meaningful Innovations Meaningful innovations aim to advance economic and social conditions of society while at the same time enhancing the competitiveness of companies. This shared value creation focuses on identifying and expanding the connections between societal and economic progress [83]. It requires collaboration between different parties, and a deep understanding of the people and their needs. For many social businesses conventional wisdom needs to be challenged and dynamic capabilities need to be developed for a company to become a stronger social force in the world, and to replace the current shareholder focus with a post-crisis system that focuses on stakeholder value [103]. Bringing together citizens, government and business provides the potential to bridge multilateral norms and local actions as a new form of collaboration to create sustainable developments [7]. In this context the value ecology metaphor of ecosystems is useful as it suggests a holistic dynamic view that emphasizes that value does not just reside in products themselves but involves the idea of a competitive and cooperative environment in which members create value in different roles [54]. The metaphor also highlights the change and dynamics in such systems.

Increasing numbers of companies are starting to realize that they are members of ecosystems, and that they are immersed in a distributed network of actors exploring future meanings. Their activities are influenced by other firms, designers, publishers, users, showrooms, events, artists, design services, suppliers, education, research and firms in other industries [97], and in their turn they influence the activities of others. A firm that takes an action without understanding its impact on the ecosystem as a whole is ignoring the reality of the networked environment in which it operates [56]. In the creative industry for example, five shifts have an impact on the business context: the shift from consumers to co-creators of value; from value chains to value networks; from product value to network value; from simple co-operation or competition to more complex 'co-opetition'; and from firm-level thinking to total value ecological system thinking [54]. The managerial challenge is to nurture the complex business communities that bring innovations to market [69]. The success of an innovating firm depends to a large extent on the efforts of other innovators in the environment. Offerings need to be combined with components and complements if they are to present value-creating solutions to customers [2]. Ecosystems allow firms to create value that no single firm could have created on its own, by combining the individual offerings of different firms into one coherent, customer-facing solution [1]. It is also recognized that solving complex problems necessitates the cooperation of all the persons concerned: non-experts and experts from different fields. Together they can identify and introduce co-operative solutions. This process differs from solving complicated problems that can be solved by an expert alone, provided he has the right information, training or expertise [63]. The current societal challenges include many such complex problems that require collaboration to find meaningful solutions.

Decision-Making The properties and modes of action in ecosystems cannot be explained from understanding the actions of the components taken in isolation. Aristotle formulated this notion as: "The whole is more than the sum of its parts". Moreover the boundaries of an ecosystem are vague because, like biological organisms, it maintains itself by a flow of molecules entering and leaving; and it is therefore difficult to tell what belongs to the system and what does not [99].

Hakansson et al. [48], emphasize that no one understands the total structure of the ecosystem. The diversity, motion and pattern of interdependence means that no individual actor can operate it alone, none can achieve change alone or predict its outcomes. Each actor has his own picture of the network, based on his own experiences, his position and relationships with others, which forms the basis for his behavior. The interconnected nature of decisions-makers in ecosystems makes defining a strategy an iterative process [2]. Companies co-evolve their capabilities around new value propositions; they work cooperatively and competitively to support new products and satisfy customer needs [69]. Business ecosystems develop through self-organization, emergence and co-evolution, in which way they acquire their adaptability. Small firms, large corporations, universities, research centers, public sector organizations and other parties can participate in, and thereby influence, the system [77]. Being involved in the ecosystem and its continuous adaptation leads to new insights and learning, and can initiate new rounds of innovations.

A common mistake managers make is to plan out the full ecosystem, pick their position within it, and act with all haste to create and defend their role in delivering an integrated product or service to the end-customer [69]. They overlook the fact that the ecosystem will emerge over time, and that all actors are interdependent. So if one experiences a problem or decides to leave the system, this will also affect the positions and opportunities of the others, for example by a delay in getting the innovation to the market. Ecosystems can also be seen as central platforms that bring actors together, in which companies try to extract ideas and use them to create new services or improve existing ones [87]. Like biological ecosystems, business networks consist of large numbers of loosely interconnected participants that depend on each other for their effectiveness and survival. If an ecosystem is healthy, the individual participants will thrive; if it is unhealthy, the participants will suffer [55].

Leadership There are various definitions and descriptions of the roles of the actors in an ecosystem or network in the literature. There is some debate about the role of a leader. Although ecosystems are not hierarchical structures, there is a need for a leadership role, especially in the development phase, to enable all members to invest towards a shared future in which they can all benefit together [69]. During the execution phase, leadership becomes a different issue. In their description of business ecosystems, Iansiti and Levien [55] identify three roles:

- Keystones: are only a small part of the biomass of the ecosystem, but exert tremendous influence. Removal of keystone species can have dramatic cascading effects on the entire ecosystem, while removal of other species, even those involved in many interactions, will have little effect beyond the loss of those connections. Because keystones themselves are dependent for their survival on the ecosystem, they will behave in a way that contributes to its health and aim to create stability in the system. Some individual members may suffer as a result of the behavior of the keystone, who may directly remove them or limit their activity, but the community as a whole benefits. Keystones can also enhance productivity from the bottom-up by providing a foundation on which the other species rely, as well as by promoting diversity. An example of a keystone in a business ecosystem is Microsoft in the computer and software industry: providing platforms and tools on which many other business partners built their businesses. Keystones leave a vast majority of the value creation to the network, and share the value they create widely, but also ensure that it captures value for itself in selected areas.
- Dominators: form a large part of the ecosystem and supplant other species in their aim to manage and control the network. They can be large in size and numbers, playing a dominant role and taking over the functions of the species they eliminate, or eliminating those functions all together. Dominated ecosystems often become unstable or vulnerable to disruption. Dominators integrate horizontally and vertically, to increase the value they can capture for themselves. They leave little opportunity for the emergence of a meaningful ecosystem. Examples are found in the early days of video recorders, where Sony aimed to control the system with Betamax, as did Philips with Video 2000, but eventually the open standard VHS was the survivor.
- Niche players: contribute to the functioning of the ecosystem with a specialized function. Niche species individually do not have a high impact on the other species, but collectively they constitute the bulk of the ecosystem. A niche player

develops or enhances specialized capabilities that differentiate it from other firms in the network. Ecosystems generally contain a lot of such niche players. They leverage the resources of the network for their own purposes, and are thereby highly dependent on the keystones. By avoiding duplication of effort, niche players contribute to the efficiency of the ecosystem and collectively create much of the value in a healthy ecosystem. They capture much of the value they create.

Leadership in ecosystems is therefore a tricky matter. On the one hand, it has proved to be helpful if a visionary leader can inspire the other members to productively collaborate to create value. On the other hand, there is the risk of a leader taking up a dominator style. So even for the leader it has to be clear that his future depends on the other members.

Interdependence A key aspect of ecosystems is the interdependence of their members. Ecosystems only function effectively when all actors who are crucial to the delivery of a product or service are healthy. Weakness in any domain can undermine the performance of the whole system. Members of an ecosystem share a common fate: they rise and fall together, as became very clear when the internet bubble imploded. This interdependency also means that companies cannot define their own strategies in isolation [55].

Members of an ecosystem are impacted by the innovation challenges experienced by its suppliers, and by its complementors [2]. If a supplier is experiencing problems this will result in the other members not being able to integrate this contribution into the offer, preventing a complete product from being presented to the customers and thereby affecting the value creation process. If the complementors experience problems, the other members are able to deliver their complete product to the customer, but the customer may not be able to use it to its full potential. This has a higher impact if the complementary offering is crucial for the functioning of the system, or more an accessory that does not directly influence the functioning of the system. For example, even if Nokia delivers its fully functional mobile phones to end-users, they will not able to use those mobile phones to their full potential if the telecom provider Vodafone has problems with its network. The value capturing process will then be severely affected. And if Jawbone is unable to deliver Bluetooth headsets, users will still be able to use their phones although perhaps not with optimum comfort, so here too there will be some effect on the value capturing process.

The long-term sustainability and stability of the ecosystem as a whole depends on aspects like these. The interdependence in the ecosystem also means that the motivations of the members should be aligned, and reciprocity in value exchange should be experienced by the members to be able to survive as individuals in the system. These are the important aspects of value at the level of ecosystems in the Value Framework (Fig. 6.1).

The role of the ecosystem in the creation and realization of meaningful innovations is to ensure that its members can keep providing value for all the stakeholders, despite inherent changes and dynamics that occur over a longer period of time. By explicitly including the level of ecosystems in value creation, this

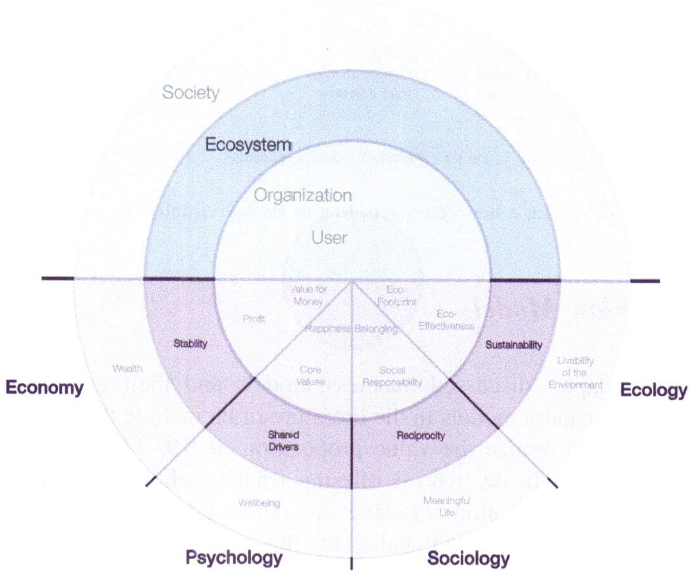

Fig. 6.1 Value elements for ecosystems in the Value Framework

ecosystem becomes the owner of 'doing good', and ensures a stable and sustainable environment in which its members can flourish. Although the ecosystem is not a single entity but a dynamic constellation of different organizations and individuals, it is able to play this role well if the common goal is aligned with the shared drivers of the members.

Designing New Ecosystems

The process of designing new ecosystems involves a number of steps. Although this will rarely be a tidy, sequential process, but is much more likely to be highly iterative, for the sake of clarity the process will be described as if it is a sequence of steps. Figure 6.2 shows the five steps, starting with the initial value proposition. This first step was discussed in the previous chapter, so this section will concentrate on the activities in the remaining steps.

Visualization is considered an important way to keep track of the development of the ecosystem, and supports all steps of the process. Before considering the steps to design the ecosystem, the Value Flow Model will be described as a method to visualize the ecosystem and to support the steps. This includes a brief description of some aspects of the Value Flow Model to clarify the model. More detailed information about these topics will be given in the description of the steps.

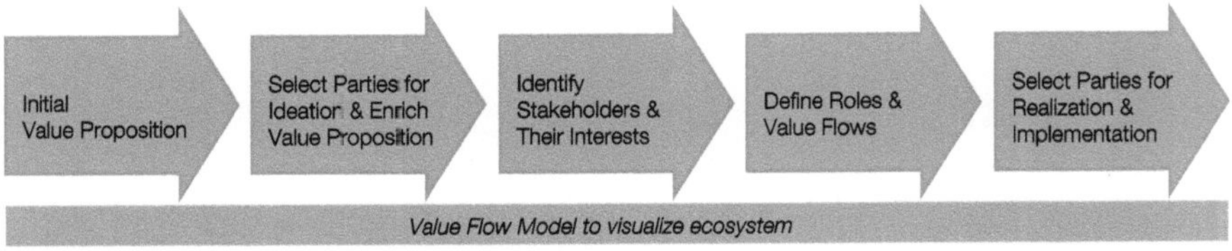

Fig. 6.2 Steps in designing a new ecosystem (iterations not visualized)

The Value Flow Model

The previous chapter discussed business models and their design parameters. Descriptions of business models in the literature often include the key partners and resources needed to realize the value proposition [9, 19, 43, 72, 104]. Business models are used to indicate 'who is offering what to whom and expects what in return' [44]. The central notion of business models is to create value for customers, and to capture a portion of that value for the organizations offering the value propositions to the customer. In the more complex situation of ecosystems or multi-party value networks, this also deals with the creation and addition of value, as well as the exchange of value between the actors. The design decisions in a business model for ecosystems include: what offerings from which actors are provided to which other actors; what are the elements of those offerings; and which value-creating or value-adding activities are performed by which actors [44]. For the more 'traditional' business models that follow the pattern of a supply chain, existing ways to shape the business model, such as the Business Model Canvas [72] are appropriate tools. While for 'new' business models requiring new ecosystems, the new Value Flow Model method has been developed. The Value Flow Model not only indicates the offerings and interactions of one organization with its customers, but also the flow of value between multiple members of the total ecosystem. The Value Flow Model combines a number of insights from socio-technical systems, value models and ecosystems and ingredients from other models as presented in the literature [3–6, 8, 9, 11, 27, 35, 41, 44, 46–48, 61, 70, 73, 89, 104], and has been enriched during use in industry projects to design ecosystems for really new and radical innovations.

A value network model visualizes specific interactions within the network to provide a perspective for understanding value-creating roles and relationships, and to offer a dynamic view of how both financial and non-financial assets are converted into negotiable forms of value. It explains how to more effectively realize value for each role, and how to utilize the tangible and intangible assets of value creation. Value models offer an abstraction that is useful in exploring new business networks and their properties, and in understanding the risks that are crucial in cross-organizational business process design. They link the customer needs and wants to the resource exchanges in the network, and provide the basis for understanding how the value is realized.

Elements of the Value Flow Model The Value Flow Model integrates different elements to provide a coherent view of how the value proposition is being created out of complementary offerings from different organizations, and how the related value flows through the ecosystem.

Actors The first element in the model is the actors. At the start of an ecosystem design project they are indicated as roles, and not as specific companies carrying out specific functions. Examples of roles are: customers (who are using or consuming the value proposition); business actors like service providers, goods providers or suppliers (who are involved in the production and sales of the value proposition); and other stakeholders like financiers or regulators (who either influence or are affected by the new value proposition). A more detailed description is given in Table 6.2 later in this chapter. The actors can be individuals, small groups or large communities.

Of course these roles cannot always be clearly identified. Users of a system may also be providers of important data for the functioning of the system (the role of a business actor). For the Genius function of Apple's iTunes, users contribute to the automatic generation of playlists by allowing iTunes to upload their personally created playlists to the system, where they are analyzed to see which songs people generally consider to fit together. Similarly, business actors are sometimes also customers of the system. The Value Flow Model can accommodate these hybrid forms, as will be shown in the section on transactions.

If different user groups are targeted, these groups are indicated separately to make sure that any differences in the offering or the underlying business models are included in the Value Flow Model. The level of detail of the model is a balance between completeness and clarity; there is no golden rule for the required level of detail. Each project team has to decide what level of detail is needed to have sensible discussions, and to make sure that the model is clear to the members of the team and stakeholders with whom the model will be discussed (Fig. 6.3).

Motivations The second element is related to the actors, and refers to their motivation: the interests and intentions of the actors or the goals they aim to achieve. The main motivations of the different actors are indicated in a very short description. This might lead to a further split of actors because of differences in their ambitions or intentions. For example some customers may focus on quality of life, while others may look for the cheapest solution. Or one manufacturer may aim to create a higher margin, while another may want to focus on volume and economy of scale. Understanding these differences is important in the design of the network. The main motivations of the actors are normally very stable, so it is not likely that they will need to be updated during the development process. But in a later stage, when specific parties are chosen for particular roles, it is very important to check the assumptions on which the ecosystem was designed. As shared drivers are an important element for the stability and sustainability of an ecosystem, it is important to choose parties that have aligned motivations. Motivation may therefore be one of the important selection criteria for the parties that will take up the roles in the ecosystem.

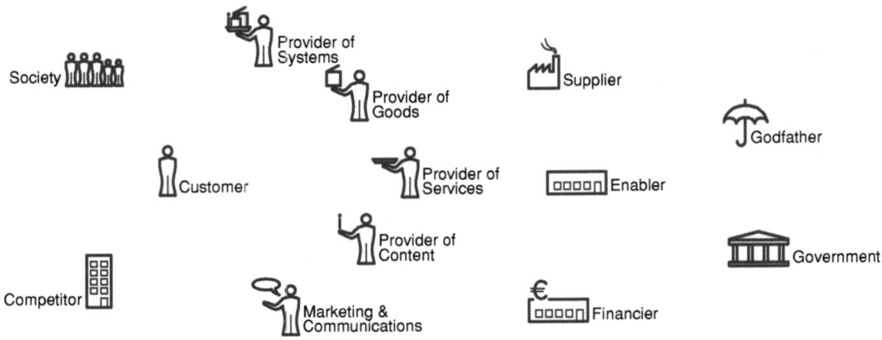

Fig. 6.3 Icons for actors (*based on roles*) in the Value Flow Model

The Value Framework, as presented in Chap. 3, is a helpful tool to discuss the individual drivers of the actors. It supports alignment of terminology and can help to broaden the discussion from economic value drivers only to include other perspectives as well. For the users, organizations and society the Value Framework indicates what the value drivers could be from the economic, psychological, sociological and ecological perspectives, as shown in Fig. 6.4.

Figure 6.5 shows some examples of motivations, and the way they are indicated in the Value Flow Model.

Compatibility and Influence The third element indicated in the Value Flow Model is the compatibility of the main motivations of the actors with the value proposition under development, and commonality of behavior and values with the initiating members of the ecosystem. For each actor, there is an indication of whether his motivation and behavior are positively compatible (+), neutral (=) or not compatible (−) with the value proposition and the initiating members of the ecosystem. A second aspect that is important to indicate is the influence of the actor on the decision-making process. This depends on his power in the ecosystem. If the new value proposition and the ecosystem are strongly dependent on this specific actor, the size of the icon indicating his compatibility increases in the Value Flow Model (relative scale of the +, = or −). There may be different opinions for different actors, and if there is a good reason for these differences, then it might be a good option to include them as separate actors in the model. Figure 6.6 shows some examples of actors and their compatibility and influence indications. In this case two organic food manufacturers with different main drivers are included in the overview: for one the main driver is profit, while for the other it is good business and social responsibility. The indication of compatibility then indicates which of the two fits best with the ambition of the new value proposition.

As the value proposition develops during the process, the compatibility with the motivations of the actors can change. The analysis can also be used to enrich the value proposition to make it more compatible with very influential actors in the system.

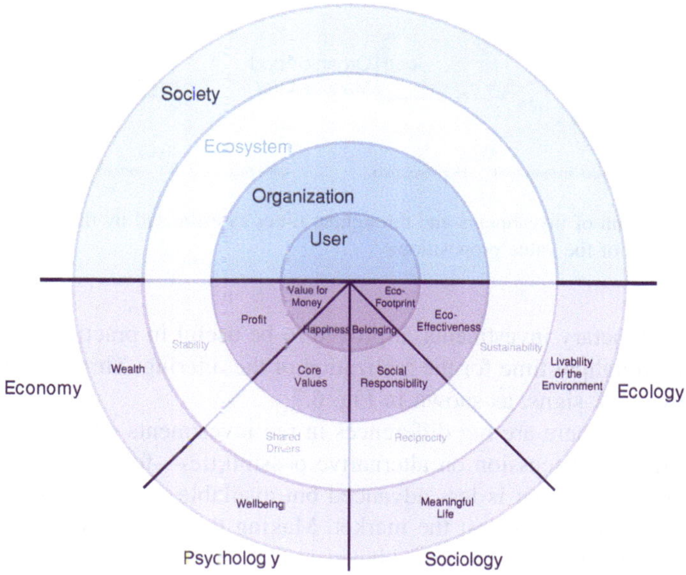

Fig. 6.4 Value for users, organizations and society in the Value Framework

Fig. 6.5 Indication of motivations for the actors in the Value Flow Model

Fig. 6.6 Indication of compatibility and influence of actor motivation with the value proposition

Investments and Throughput Time For the different business actors, it is useful to make a rough indication of the investments required to develop and realize their parts of the value proposition. Ballpark figures for the investment are sufficient, as the model is not intended to make a detailed business case of the complete offering and all investments per company. An understanding of where the big investments need to be made will help to make sure there is a balance between investment and revenue streams later on. In the Value Flow Model the investment level is indicated by euro signs as shown in Fig. 6.7.

Fig. 6.7 Indication of investments and throughput times as estimated by the business actors to deliver their part of the value proposition

Next to monetary investments, it proved to be useful in practice to indicate an estimated throughput time for the realization of the offering. Throughput times are indicted by clock signs, as shown in Fig. 6.7.

Especially, if there are big differences in the investments or throughput times, this can trigger a discussion on alternative possibilities—for example whether to have a market offer that is less advanced but available earlier, or to temporarily accept a lower margin to test the market. Making these choices visible will also trigger the discussion if the potential business partners are willing to share the risks and investments. Having these discussions openly early in the project helps to build trust between the members, but also enables the uncertainties to be reduced as early as possible in the planning process.

Transactions Transactions form the fifth element of the Value Flow Model. Transactions can be activities that originate with one actor and end with another, or resources, information or items that are shared or exchanged between two actors. Transactions are indicated by arrows that also show the direction of the flow, with labels to show the content of the transaction (Fig. 6.8). This content can be of different types:

- Goods and services: physical goods are the most visible when moving around, but services can also be part of the value proposition offered to the customers. If you are enjoying dinner in a restaurant, you are not just offered the food itself, but also the services of preparing and serving the food. Not all the details are shown in the Value Flow Model for the sake of simplicity, but in the example the transaction can just be indicated as 'dinner'. To provide the dinner the restaurant needs to obtain the ingredients from suppliers, which is indicated by goods flows towards the restaurant from various supplying business actors. In this case the restaurant has goods flowing in (arrows pointing towards it), and goods flowing out (arrows pointing away).
- Money and credits: money involves both the revenue streams that start with the payment of customers for the value proposition, and the monetary transactions that share these revenues between the business actors, and also loans or credits. Credits can be monetary but also virtual, such as 'air miles' or loyalty cards. The monetary flows cover actual payments, and also contracts and invoices.
- Information: one actor can share information with another in different forms: knowledge, advice, information about something relevant for the value

Fig. 6.8 Indication of the different types of transactions in the Value Flow Model

proposition, or raw data. Typical information can be strategic information, technical knowledge, user data, planning information etc. Information is an increasingly important aspect of innovation, especially with the rise of the information and communication technologies that enable all kinds of intelligent systems. Some information transactions could also be labeled as 'goods', such as when knowledge is shared through physical documents. Since information plays such an important role, it is indicated as a separate flow in the Value Flow Model. Especially, if the ambition is to create a sustainable change, the monitoring of the progress of the change is a very important aspect that requires dedicated information flows. Knowledge generation and knowledge application are typically important activities that actors in the ecosystem do to create value through information.

- Intangibles: intangible values are benefits or assets like experience, reputation, exposure, attention, increased well-being. A characteristic of intangible values is that they do not change from ownership in the transaction.

For a sound model there is reciprocity in the transactions, either direct or indirect, through exchanges of value with other actors. The difficulty in understanding the balance lies in the fact that the different types of transactions are not directly comparable.

Core Value Proposition To increase the clarity of the overall Value Flow Model the core value proposition will be highlighted as compared with planned or potential complementary offerings and the supplying and enabling network. The core value proposition includes the end-customer, users and those actors who exchange value directly with the end-customers or users. Without the elements in the core value proposition, there is no value at all for the customers or users.

What is actually part of the core value proposition also depends on the level of integration in the way the new proposition is presented to the customer. Figure 6.9 shows an example of three actors together providing the core value proposition of healthy eating for families with small children. All three actors have direct access to the customer, each providing a part of the value proposition. For the same value proposition, an alternative scenario could be that only the healthy eating service is presented as the core value proposition, and the other actors are mainly suppliers and enablers without direct access to the customers, as shown in Fig. 6.10.

Complementary Offerings Next to the core value proposition, the complementary offerings are indicated in the Value Flow Model. The complementary offerings can make the total bundle of offerings more attractive for the customers, and

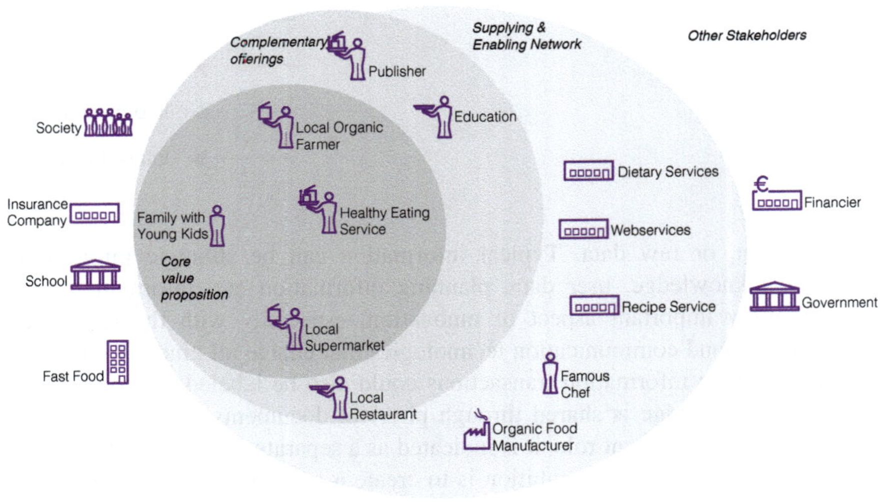

Fig. 6.9 The Value Flow Model: the core value proposition at the center, surrounded by complementary offerings, supplying and enabling network, and the other stakeholders at the periphery

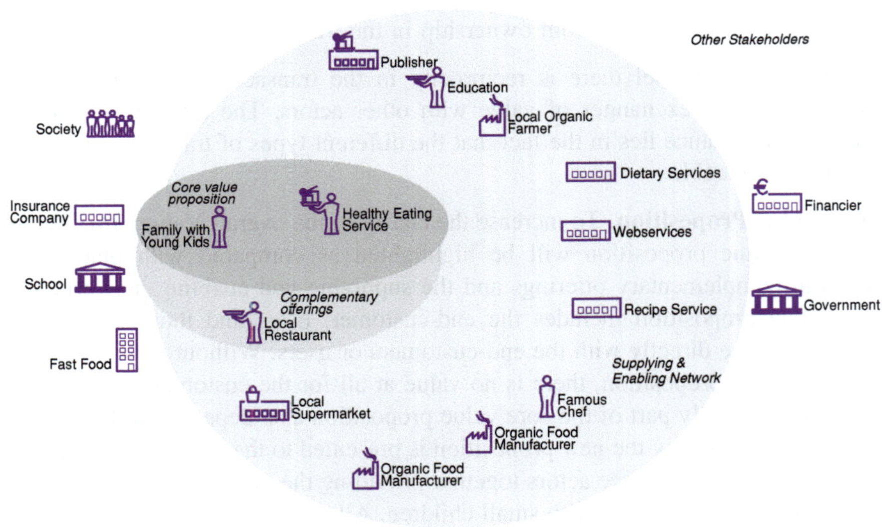

Fig. 6.10 Alternative scenario of the same value proposition, with a more controlled interface towards the customer and more actors in supplying or enabling roles

they may also address additional target groups. The actors involved in the complementary offerings also have direct contact with the customers or users, but they are not core to the system. The value proposition can still work without these

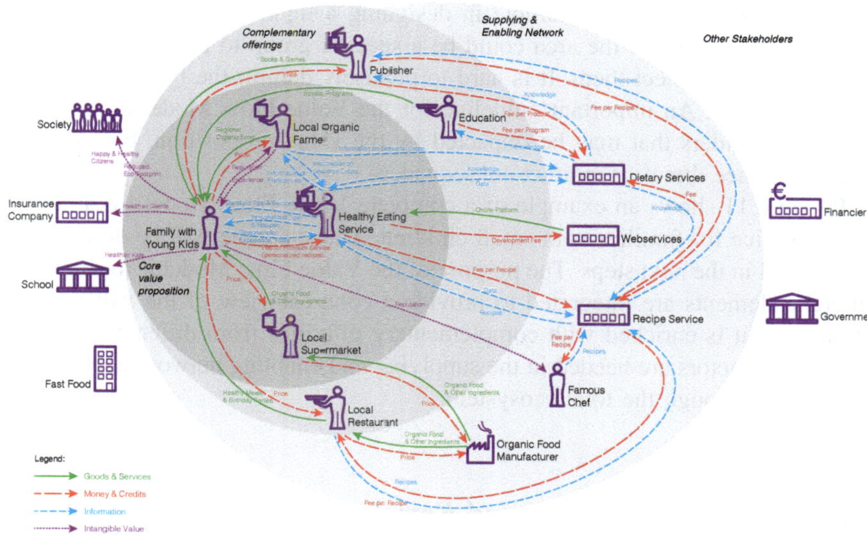

Fig. 6.11 Example of a Value Flow Model for a healthy eating service

complements. In the value network of Apple's iPad, for example, complementary offerings are the accessories of the iPad, such as the covers, connectors or dedicated apps. The iPad fully functions without these complementaries, so they are not considered core to the value proposition. But for the user they certainly increase the value of the total offering. In this example the iPad is part of the core offering, as without it there is nothing to offer to the customers, but iTunes is also part of the core offering, as an essential part of getting the iPad to work. The same applies to the operating software of the iPad.

Supplying and Enabling Network The supplying and enabling network includes those actors and transactions that deliver components for integration into the value proposition (such as suppliers of hardware components) or that play a role in enabling the value proposition (enablers like platform providers or regulatory bodies). There may be several alternative scenarios during the development of the ecosystem, so in the first iterations the roles are indicated, and other relevant aspects such as key technologies can also be mentioned. But the actual organizations or companies that can provide the product, service, knowledge or intangible asset component of the service proposition will only be selected later.

Other Stakeholders The last element that is indicated in the Value Flow Model is the stakeholders that are affected by the value proposition but are not directly involved in it. It is important to include these stakeholders so that the new value proposition does not create (unintended and/or unwanted) externalities. Society at large is typically an actor that should be included in any project, but depending on the project more specific stakeholders may be identified that could be impacted by the value proposition. For example in designing a lighting concept for a public space, the neighbors of the area could be a specific group to be identified, as the stray light could affect them. It is hard to precisely delimit the boundaries of the total ecosystem. An important guideline for the Value Flow Model is to include those stakeholders that may be impacted by the value proposition, because they will experience direct or indirect consequences.

Figure 6.11 shows an example of a complete Value Flow Model for a healthy eating service for families with small children. This example will be described in more detail in the next steps. The purpose of the Value Flow Model is to show how different elements are integrated to provide a coherent view of the value proposition, how it is enriched with complementary offerings from different organizations, which actors are needed in the supplying and enabling network and how the value flows through the total ecosystem.

Select Parties for Ideation and Enrich the Value Proposition

In this chapter it is assumed that an initial value proposition and an accompanying business model are available, as described in Chap. 5. The first step in the process of designing an ecosystem is then to select parties for further ideation, and to enrich the value proposition. For the sake of clarity, the design team of the ecosystem is referred to as project team; the members of the ecosystem under development are referred to as ecosystem members. It could well be that some of the design team members will also become ecosystem members, but that is not necessarily the case for all of them. The project team is not a fixed team either; depending on their knowledge and experience participants may leave and enter at any stage. In this phase there are usually no contracts (apart from non-disclosure agreements), and participants contribute without being sure that they will get any return on investment. Experience in such projects shows that people participate because they believe in the opportunity, without yet knowing exactly how they will be able to benefit from their contribution.

This step starts with an understanding of which parties could contribute with relevant knowledge of the users, the market, the business system, technology options or regulatory arrangements. These parties are invited to take part in a process of collective exploration [28]. The mindset of the participants in this process should be flexible enough to prevent a lock-in of certain parties, and allow for the inclusion of alternative scenarios or ideas. Together the project team explores the potential of the initial value proposition, and enriches it with ideas that could further improve its value. The main activity is to identify the key resources needed to deliver the value proposition to the customer, including: people, technology, products, equipment, information, channels, partnerships, alliances or brands [58].

Designers can play an important role as moderators in this process to help bridge the 'symmetry of ignorance' [59]. In this notion it is understood that

Table 6.1 Roles in the project team for the development of the ecosystem

Role	Description
Creator	The architect, or guru of the project, provides the overall vision and guidance to the evolving team, inspires other participants, triggers and catalyzes innovation, directs innovation activities and tends to the network
Knowledge expert	Contributes to the project with ultimate sources of explicit knowledge, adapts to the direction set by the visionaries, and contributes by providing specialized knowledge or service and infrastructure related to his expertise field
Resource provider	Provides critical resources needed in the iterative innovation process
Collaborator	Adapts to the direction of the architect, playing a supporting role by coordinating and organizing tasks such as design, build and test activities; carries out supply chain-related activities such as organizing pilot production, distribution; or marketing and business planning activities
Communicator	Ambassador of the project, with links to external networks, 'connector' who supports the project by bringing in relevant participants to workshops or new potential project members. Serves as a broker or bridge in the network, linking members and mediating knowledge transfer
Moderator	Provides structure in the innovation process, supporting the project by ensuring all important members have a common understanding of the goal of the project, and can make their maximum contribution; facilitates problem-solving, and guards the culture of the of collaboration

nobody has all the required knowledge. For example users are domain experts on their practice, and know implicitly what the value proposition should achieve; while system developers know the technology, and know how to achieve a system that can do it. In this phase narratives can play a role to support the discussion between the participants. Narratives enable the recombination of ideas, and link present innovative ideas with past experience and future aspirations [10]. The moderator of the process can trigger the participants to share stories and explain visions in a more detailed way. More 'words' are needed to create a shared understanding, especially when people of different disciplines and with different experience and backgrounds collaborate. There are sure to be differences in culture and profession, in rationality and logic, and in the perception of the objectives, criteria and constraints. It is therefore important to work on a common language, and to increase the number of viewpoints so that all participants understand the value proposition that is emerging from the different perspectives [63].

Participants may play a number of roles in the development of the ecosystem [42, 46, 51, 65, 67, 71, 92, 94, 98]. Table 6.1 indicates some common roles in the ecosystem development project team

The iterative process involves gaining a better understanding of how the value proposition would address the insights and needs of the targeted users, as well as finding solutions to provide such a service proposition so that the value is delivered in a way that is attractive for those users. Along the way, new insights could be generated that require new parties to join the process and make a further iteration.

During the process the Value Flow Model helps the discussion and triggers the options for enrichment of the value proposition. The aim in this phase is to get the core value proposition right, and to address valuable complementary offers as well. The aim is either to make the value proposition more interesting for the targeted users or customers, or to be able to increase the market potential by addressing additional target groups.

In the process of building the project team by integrating relevant new participants who may also play a role in the business ecosystem itself later on, it is important that the partners can create synergistic value. This requires alignment at three levels [34]:

- Knowledge level: complementarity of both technical and market knowledge, to ensure a mutual understanding of the market opportunities, and the capabilities of technologies and their implications in the market;
- Strategic level: correspondence of the goal of the project, which enables the development of a team for the co-development process;
- Relation level: compatibility of cultures, ways of working, willingness to change and long-term orientation.

A team will only function well if the members' expectations of each other are clear. It can be a helpful step to actively ask all the members what they expect from their own involvement in the system, and what they expect from the others [70]. This will highlight incompatibilities at the knowledge, strategic and relation levels early in the process, when there is still time to adapt or to find new partners to fill in gaps. Of course the team will experience uncertainties in the collaboration, so it needs to have a good understanding of [100]: who are the parties involved; why are they participating, and what do they want to achieve; what are they specifically interested in; which way do they want to achieve the goal and what needs to be done; what resources are involved or required; and when do they expect the goal to be achieved. As most organizations behave opportunistically, the team needs to ensure that it installs win–win situations within the developing network. To provide network stability and flexibility at the same time, all partners have to agree on the aims of the innovation project in advance and clarify where and when the partnership starts and ends for them [91].

Once the team is confident that the purpose of the project in terms of the core value proposition is sufficiently clear, and that the team itself is a strong team, it can move to the next step.

Identify Stakeholders and Their Interests

Freeman's classical definition of a stakeholder is: any group or individual who can affect or is affected by achievement of the firm's purpose [38]. Or in other words, those who experience or anticipate experiencing potential benefits or disbenefits as a result of an organization's actions [29]. In stakeholder theory, it is considered the manager's responsibility to act in the interest of stakeholders, to maintain and take care of relationships and networks of stakeholders, and to act to care for the earth [38]. The theory rejects the separation of ethics and economics; it assumes that values are necessarily and explicitly part of doing business by asking managers to articulate the shared sense of the value they create for its core stakeholders [39]. The stakeholder approach includes critical and often neglected aspects of the social and political environment, and emphasizes the notion of an 'extended enterprise' that is engaged in diverse functions and has links with numerous other entities and individuals, who can be potential beneficiaries and/or risk bearers [84]. Stakeholders can be found in the internal activity system of an organization and in its external ecosystem [32]. The analysis in this chapter does not include the internal stakeholders within the actors boundaries. The focus here is on the external stakeholders in the ecosystem, who are involved in or might be experiencing benefits or disbenefits of the new value proposition under development. The literature on stakeholders usually focuses on a firm or corporation, and identifies stakeholders for that specific organization. In the context of this book this definition needs to be modified, as the starting point is not the firm but a value proposition for an innovation that might be brought to the market by a network of organizations. The stakeholders are therefore any groups or individuals who can affect or are affected by the achievement of the innovation.

Identifying Stakeholders

There are several approaches to identifying stakeholders. Each of these takes a slightly different view, and might inspire a team to identify a specific actor who needs to be included in the Value Flow Model. The stakeholder model originally used rather large stakeholder groups such as customers, suppliers and investors, and ignored many of the differences within such stakeholder groups [53]. A more fine-grained view is needed, but stakeholders with similar interests can still be classified as belonging to the same group [20].

Primary and Secondary Stakeholders Primary stakeholders are those whose continuing participation is necessary for the survival of the core value proposition as a going concern. There is a high level of interdependence between the core of the ecosystem and its primary stakeholder groups. If any of them becomes dissatisfied and withdraws from the ecosystem, in whole or in part, the ecosystem will be seriously damaged or unable to continue its activities [20]. Examples of primary stakeholders are: customers (from different customer segments), suppliers (different firms), financiers (shareholders, investment banks, commercial banks) and communities [38, 40].

Secondary stakeholders are defined as those who influence or affect, or are influenced or affected by, the innovation, but are not engaged in transactions with the ecosystem and are not essential to its survival. The ecosystem is not dependent on such groups, but they can cause significant damage. The media and a wide range of interest groups are considered as secondary stakeholders under this definition. They have the capacity to mobilize public opinion in favor of or against the innovation [20]. Examples of secondary stakeholders are: competitors (domestic or foreign), consumer advocate groups (consumer federation, consumer's union, council of consumers), special interest groups (e.g. safety and health groups, environmental groups, single issue groups), media, trade associations (business round table, customer trade organizations), political groups (political parties, national league of cities, national council of mayors) and government (congress, courts, cabinet departments) [38, 40].

Six Markets Model Stakeholders can also be identified through the six markets model [76]. The six markets that the model refers to are:

- Customer markets: existing and prospective customers and intermediaries;
- Referral markets: existing customers who recommend their suppliers to others and referral sources or 'multipliers', such as accountants who refer their customers to a law firm;
- Influencer markets: financial analysts, shareholders, the business press, government and customer groups;
- Employee markets: attracting the right employees to the organization (not relevant in this context);
- Supplier markets: traditional suppliers and strategic alliances;
- Internal markets: internal departments and staff (not relevant in this context).

Sociopolitical Stakeholders A sociopolitical analysis distinguishes four types of actors [94]:

- Beneficiaries: people or parties who will benefit from a projects success, but might be suspicious;
- Potential allies: who may be willing to commit support to the project, such as influential members of government or a potential godfather for the project;
- Indifferents: who have no strong opinion on the project, but whose support, effort or resources may be necessary for the project's success;
- Meaningful opponents: who will be adversely affected by the project's success and may resist or delay its execution.

The Harm Chain Analysis of the new value proposition from the perspective that it causes undesirable impacts on stakeholders. The analysis identifies: those causing harm; those being harmed; and those regulating harm. Its aim is to reduce the undesirable impacts on stakeholders, and to reduce the 'gaps' in expectations [81].

Examples of Stakeholders To illustrate which stakeholders were deemed relevant in projects, the example of a case of looking for innovations in the context of heart attack patients is used. The identified stakeholders in this project are: the patient (child, young adult, woman, man, senior), general practitioner, neurologist, radiology staff (nurse), interventionalist, insurance company, intensive care staff, rehabilitation specialists, rehabilitation physician, health center, home

care, social worker, county council (for benefits), INR staff, bystander, emergency services telephone operator, emergency services personnel (ambulance driver), emergency department physician, family/partner, emergency department staff (nurse), stroke unit staff (nurse), pharmacist, neurosurgeon, intensive care physician, employer [75].
The Value Flow Model is used to show the relevant stakeholders as actors on the map, taking into account whether they contribute to the core value proposition (primary stakeholders), provide complementary offerings, supply or enable offerings, or are other stakeholders on the periphery of the new value proposition. The result is a first rudimentary version of the Value Flow Model. Figure 6.12 shows an example of stakeholders for a healthy eating service.
Positioning the actors in the Value Flow Model can trigger a discussion on potential inclusion of the stakeholders in the proposition. For example, the insurance company may have an interest in health, especially in the costs related to unhealthy lifestyles of children, and the team could decide to invite an insurance company to become a project member. Similarly, the school could also have an interest.

Analyzing Stakeholders and Their Characteristics

Once the relevant stakeholders for the new value proposition have been identified, the next step is to indicate their interests and characteristics. As discussed earlier in the Value Flow Model, three aspects of the actors are indicated: their motivation, the compatibility of their motivation and behavior with the new value proposition, and an indication of the investments they will need to take to realize the value proposition together with the corresponding risks.

Some additional analyses can be carried out to gain a good understanding of the stakeholders, their stakes and the influence they might have on the value proposition and the emerging ecosystem.

Stakes A first aspect to consider in more depth is to understand the stakes of the different actors. Some examples of stakes are [49]:

- Financiers: have a financial stake, having invested in the new value proposition with the expectation that their investment will grow in value and/or will bring income; some may expect the ecosystem to meet specific standards;
- Customers: expect benefits from the products and services they purchase; they tend to remain loyal as long as their needs are satisfied at a reasonable price; they may have an interest in the survival of the ecosystem if they cannot get the products or services they need conveniently from another company at a comparable price; they also have a stake in the reputation of the company;
- Suppliers: have a stake in the well-being of the ecosystem for continued business, and for a steady source of sales revenues;
- Society at large: the local community, and to some extent the people of the state in which the ecosystem operates; their interests lies in the tax revenues, jobs,

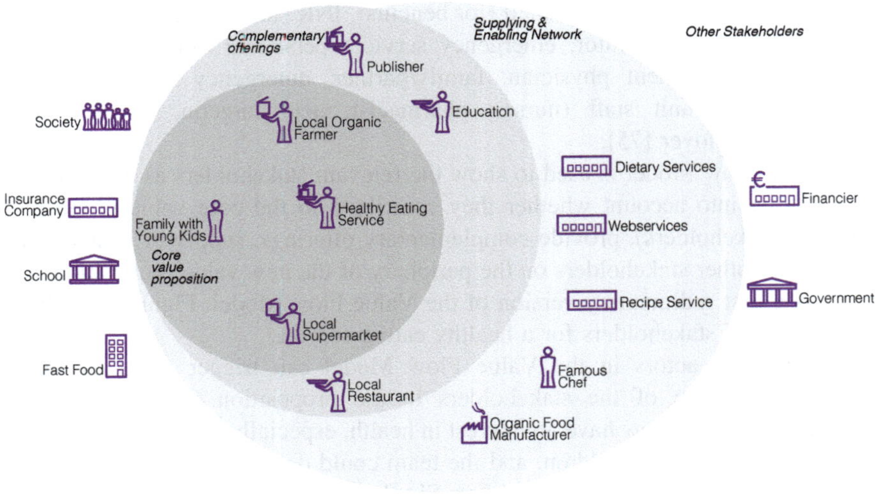

Fig. 6.12 Example of stakeholders for a healthy eating service

volunteer activities, charitable contributions etc.; they also have a stake if the new value proposition has an impact on the environment, e.g. through pollution, traffic or quality of life (in a positive or negative sense).

Some of the actors who have been identified as stakeholders, such as the media, may not have a stake in the value proposition itself, but can be powerful influencers [29], so it is important to understand their intent and motivation.

Power and Influence The next important aspect to understand is the power and influence of the different actors. Some stakeholders will have more power than others. Their power can also be of a different nature: some may have economic power, others political [38]. For each stakeholder the relative power—or ability to influence decisions in the ecosystem—can be indicated. Stakeholder power is often difficult to define, but not difficult to recognize. Powerful stakeholders are able to bring about the outcomes they desire [18]. This may be due to having many ties to other stakeholders, or being able to control other stakeholders (like the dominator type of role discussed earlier), or having economic or political power.

Another factor that is important in this respect is the legitimacy of an actor: to what extent is he accepted within the network of relationships. Legitimacy is the perception that the actions of the stakeholder are desirable, proper or appropriate within some socially constructed system of norms, values, beliefs and definitions [18]. The legitimacy of the total ecosystem can be increased if there is a balanced representation of the stakeholders, and if accountability and transparency are assured in the selection process of representatives for stakeholder groups [7]. The legitimacy of the actors in the emerging ecosystem is often only very subjectively addressed, and the notion of the ecosystem 'culture' is growing over time.

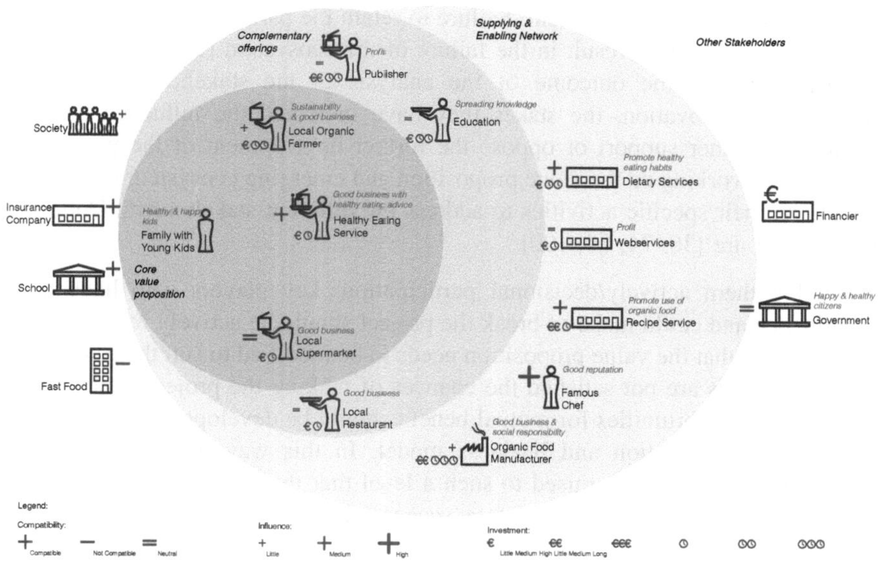

Fig. 6.13 Example of motivation, compatibility and influence of stakeholders

Position and Predictability of Behavior Different stakeholders may have different positions towards the project: they may be opposed, not committed or supportive of the new value proposition. In this context it is also important if the team can predict the likely behavior of the stakeholder. Combining these elements, there are five levels: active opposition, passive opposition, not committed, passive support and active support [100].
The stakeholder analysis is used to generate knowledge about the relevant actors so as to understand their behavior, intentions, interrelationships, agendas, interests and influences, and their impact on the decision-making process [16]. Taking into account their stakes, power influence and positions towards the new value proposition, the team can define a strategy for actors who are not yet actively involved in the project (Fig. 6.13).

Define Stakeholder Management Strategies

Just as it is the responsibility of management to select activities and direct resources to obtain benefits for legitimate stakeholders [29], it is the responsibility of the emerging ecosystem to ensure that the relevant stakeholders are committed to the success of the new value proposition. The survival of the ecosystem and its continuing success depend on its ability to create sufficient wealth, value or satisfaction for those who belong to each stakeholder group, so that each group continues as a part of the system. Failure to retain the participation of the primary stakeholder group will result in the failure of the ecosystem [20].

Depending on the outcome of the analysis of the stakeholders' positions towards the innovation, the stakes they have in it and the influence they can exercise to either support or oppose the further development of the project, the team that is working on the value proposition and emerging ecosystem will have to decide on their specific activities to address the different stakeholders. Options in this respect are [38, 74, 84, 100]:

- Involve them actively/decisional participation: key players who have a big influence and could make or break the project should be actively involved. This may mean that the value proposition needs to be modified to suit their needs, but if their needs are not satisfied the chances of making the project a success are very low. Opportunities for mutual benefit should be developed and included in the value proposition and business model. In this way the interests of the stakeholders should be raised to such a level that they see the value of participating in the project to make it a success. These stakeholders then become active participants in the decision-making process.
- Keep satisfied/consultative participation: actors who have a big influence but little interest in the project should be kept satisfied, to prevent them from becoming opposed to the project: if their needs are not satisfied they may use their power to break the project at a later stage. They do not need to be actively involved in decisions, but the team should keep track of the interests of these actors by constantly monitoring them and asking them for their views and perspectives on specific issues or plans at a deeper and more explorative level.
- Keep informed/informative participation: actors with high interest but little influence should be kept informed, although their values and attitudes are not further explored.
- Minimal effort: actors who have low interest and low power do not need a lot of attention. It is nevertheless wise to keep them on the map, and to keep their existence in mind.

Based on the outcome, the team that is working on the value proposition should actively approach the stakeholders who are key players and involve them in the project. When approaching a stakeholder it is important to prepare the discussion by adapting the narrative and the calculations that support the value proposition and the business model towards the specific situation of the stakeholder [28].

Involving new members in the process of developing the ecosystem may result in a redefinition of the value proposition and the business model, as well as an update of the Value Flow Model, because a new member of the emerging ecosystem may bring in new insights, different perspectives or new resources and ambitions.

In the example of the healthy eating service, the team could decide to invite the insurance company to become actively involved in the project. In preparing the pitch the team will then discuss what makes the proposition attractive to an insurance company. This could very well include proof of the actual health gains of the proposition, and the cost savings in caring and curing. This could involve the inclusion of a doctor in the discussions to ensure that actual health gains are registered. The proposition could be enriched by collecting data on the actual medical costs of the users. Such discussions obviously involve ethical dilemmas on the extent to which a value proposition can use such data so as not to invite 'big brother' criticisms. But this might be perceived differently if the insurance company decided to give a discount on the premium service in return for data from customers.

Define Roles and Value Flows

The next step in the process is to further design the ecosystem by defining the roles of the actors in the network and designing the value flows between them. In this step some important decisions will be taken that will have an impact on the next step—the selection of parties that will actually execute the steps to realize and implement the value proposition, launch it on the market and start doing business with it. Experience in industry projects shows that it is important not to select parties for these roles too early in the process, as this reduces flexibility in creating the best value model for the customer and in creating a win–win situation for all actors involved. So in this step, the actors and value flows are indicated without already deciding which organization will indeed be playing this role.

Define Roles for the Actors

Some roles in the ecosystem may seem obvious: customers and users of the value proposition are the center of the model, the target of the whole innovation process. But they also can be involved as co-creators, and contribute to the service.
For the business actors in the model there are various roles that need to be played to get the total ecosystem working. The combined list of roles shown in Fig. 6.14 is based on different descriptions in the literature [9, 11, 21, 24, 25, 93, 94, 101, 102], and on experience in different industry projects.

The important thing in designing ecosystems for really new or radical innovations is to keep the customer central in all discussions and decisions. This has proved to be the only way to maintain alignment of the members of the collaborating team, all of whom come from different organizations with different views and ambitions. Design choices in the definition of the value proposition and business model have the ultimate aim of delivering sufficient value for the customers and the network of collaborating actors [46]. By first of all focusing on the value for the customers, the team can maximize that value. As described before, this means making the cake as big as possible before starting to divide it.

Some critical design decisions will need to be taken in defining the roles [5, 9, 26, 46, 80, 82]:

Role	Description
Customer	Users, buyers, co-creators: the target of the new value proposition who receive the value but also may contribute by co-creating or delivering information
Intermediary	Business actors like retailers, brokers who have direct contact with customers in selling products and services of other providers, they may provide a context and lead the choreography of the contribution of the other business actors
Provider of Systems	Business actors integrating different goods and services (their own and from other providers) into a complete system and providing those directly to the customers
Provider of Goods	Business actors providing physical goods directly to the customers, either as part of the core value proposition or complementary goods
Provider of Services	Business actors providing specific services in the value proposition or generic services, such as billing, financial transaction management, logistics and delivery, customer care or service management in direct contact with customers, either as part of the core value proposition or complementary services
Provider of Content	Business actors creating content and providing it directly to customers
Supplier	Delivering components (hardware or services) to providers of goods or systems that will be integrated and delivered to the customers. Suppliers do not have direct contact with the end-customers.
Enabler	Delivering goods or services to the providers that help enable the delivery of the service, but is not part of the offer to the customer (e.g. providing an infrastructure e.g. an online platform, arranging security and privacy, approbation). Auxiliary enablers, such as e.g. utilities, are also essential to the entire ecosystem, but are not included in the Value Flow Model.
Financier	Investors, who enable the development and implementation of the new value proposition with financial support
Marketing & Communications	Business actor who is dedicated to the commercial communications and marketing activities to promote the new value proposition
Godfather	Person with political influence who can shield the project from undesirable intervention, especially when the project is scaling to a size where it could disrupt or throw unwelcome light in prevailing socioeconomic or sociopolitical realities
Competitor	Competitors can be important allies in building legitimacy, creating demand and supply of new technologies in radical innovations.

Fig. 6.14 Roles in the business ecosystem

- Customer ownership: Which actor(s) will have direct contact with the customers? Will there be an intermediate role, or will the providers of systems, goods and services directly access the customers? If more than one actor is facing the customer directly, how will the customer experience be orchestrated?
- Customer relationship: Will the service proposition be mainly transactional, or does it involve a continued relationship? Which actor will maintain the relationship with the customers?
- Network openness: Who is allowed to join the network? Will it be open for everybody to participate, or is it a walled garden? If the network is not open, who decides if new parties with new, complementary products or services can join?
- Collaboration with competitors: To what extent is collaboration with competitors necessary or desirable to build legitimacy for the innovation, or to create the market?

It may be useful to discuss a number of different scenarios for the roles and related decisions, and to explore the consequences of the various decisions. These decisions are reflected in the Value Flow Model by positioning the different actors in the two-dimensional space.

In the example of the project to develop a service to improve the eating habits of families with small children, the aim is to increase the use of regional, seasonal and organic food, and to make people more aware of healthy and tasty meals through edutainment. The business model consists of a free standard service that provides some general information and standard recipes through an Internet site, and a premium service that provides personalized recipes based on taste, dietary information and feedback on earlier recipes. Two scenarios for the business model of the premium service could be considered: direct exchange, in which customers pay separately for each item they buy, or a monthly subscription that includes a whole package of recipes, ingredients, books, computer games and yearly birthday parties at a local organic restaurant. These scenarios differ not only in the business model offered to the customers, but also in who is actually involved in the direct customer relationship. In the first scenario (also shown in Fig. 6.9), all the actors offering part of the service have direct contact with the customer. In the second scenario (also shown in Fig. 6.10), only the healthy eating service provider and the restaurant organizing the birthday parties have direct contact. However, bookings for the restaurant are not done directly but through the healthy eating service. Figure 6.15 shows the two scenarios of the core and supplementary value proposition in the Value Flow Model.

Defining the Value Flows

The next activity is to define the value flows among the actors: indicating the transactions that take place between the customers and the different actors. These transactions cover goods and services, money and credit, information and

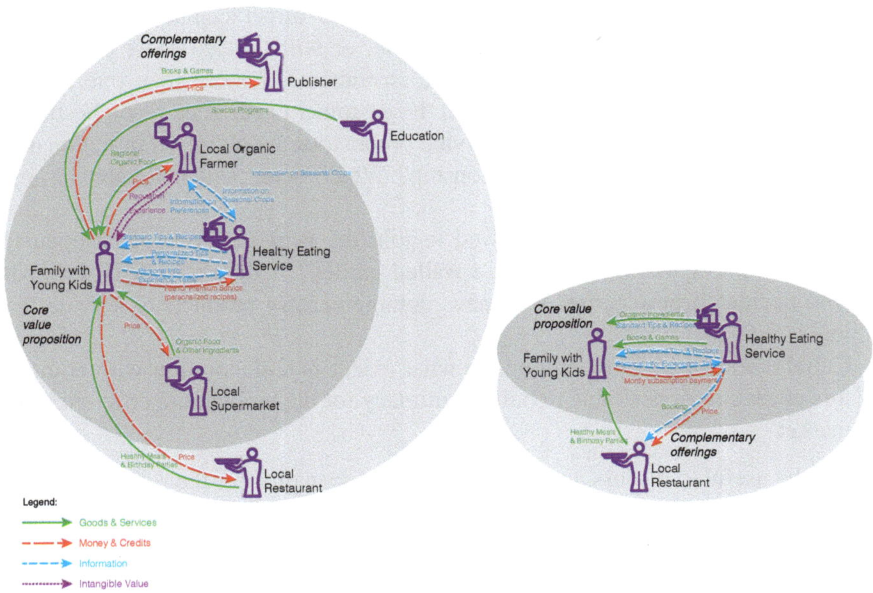

Fig. 6.15 Two scenarios for the healthy eating service proposition showing the different actors providing the core and complementary services

intangibles. Defining the transactions usually inspires creativity in the team, through which new opportunities to create value are identified. Of course such creativity is encouraged in the iterative process of developing innovative value propositions.

In industry projects it has proved to be useful to indicate these different options in scenarios and to use them for a discussion on priorities in the market introduction of the different components of the total system. The results can then be used to create a roadmap to plan extensions to the core value proposition over time. Such a roadmap contributes to the overall strength of the ecosystem, but keeps the initial steps manageable because not all components and activities need to be executed at the same time.

In defining the value flows, some critical design decisions will again need to be taken [9, 22, 37]:

- Actor relationships: Will the interactions between the actors in the ecosystem be based on relationships or will they be mainly transactional? How do we ensure reciprocity between the actors?
- Cost sharing: Will the investments and costs of the total ecosystem be controlled centrally and shared among the participants, or will there be a distributed system in which each actor needs to take care of his own balance in value?
- Revenue sharing: How will the revenues be shared among the members of the ecosystem?

The resulting transactions are all indicated in the Value Flow Model. Figure 6.16 shows the result of a discussion on the transactions in the healthy eating service.

Designing the Value Flow Model is a highly iterative process, which includes continuous checks and discussions with different members and potential stakeholders to create a picture that adequately represents the value proposition, the value flows and the actors in the ecosystem who will eventually all play a role when the proposition makes it to the market. In practice, several scenarios are usually explored and checked on their attractiveness and feasibility.

Analyzing and Improving the Value Flow Model

During the process to design the ecosystem a number of checks are useful to further improve the model. These checks include consistency, coherence and level of completeness but also specific analyses related to the value flows in the system.

Completeness and Sequence A first check of the model is about the completeness and the sequence of the delivery of the goods and services. In this analysis it is assumed that the business model of the initial value proposition is sound, as discussed in this chapter. Starting from the customers or users of the system, a check is made of whether all the elements of the value proposition that is offered to the customers are included in the model. This includes the physical goods and services, but also the sequence of activities needed by the various levels of actors supplying to each other. Following the reverse flow of elements of goods and services, the model is checked and improved where necessary.
The example in Fig. 6.17 shows the goods and services related to the healthy eating service. Although the core of the proposition is the system provided by the healthy eating service, this actor is not involved in goods transactions. The service is built on information, as shown in more detail in Fig. 6.19.
The financial flows are shown in Fig. 6.18. The standard service is free, and includes standard tips and recipes. The premium service is available for a fee. The business models in this example are rather straightforward: there is a direct payment for the services and goods obtained, based on fees and prices. This direct exchange takes place between the various providers of goods and services and the customer, but also between the providers and their suppliers.
The premium service includes personalized tips and recipes provided by a recipe service, and builds on the knowledge of a dietary service. Specific conditions and personal preferences are taken into account in an effort to make the experience of eating local, organically grown food a pleasant experience for children. A first indication of the information flows is given in Fig. 6.19. As the proposition builds on a premium service that provides personal advice to the customer, specific customer information is needed, such as food preferences, appreciation of the recipes provided and specific dietary needs. Next to this information, the service also aims to promote the use of regionally grown organic food, hence up-to-date information about the availability of these crops and meats is also needed. At this stage of the flow, the team could have a discussion on whether they would like the information about availability to be really up-to-date, perhaps even indicating

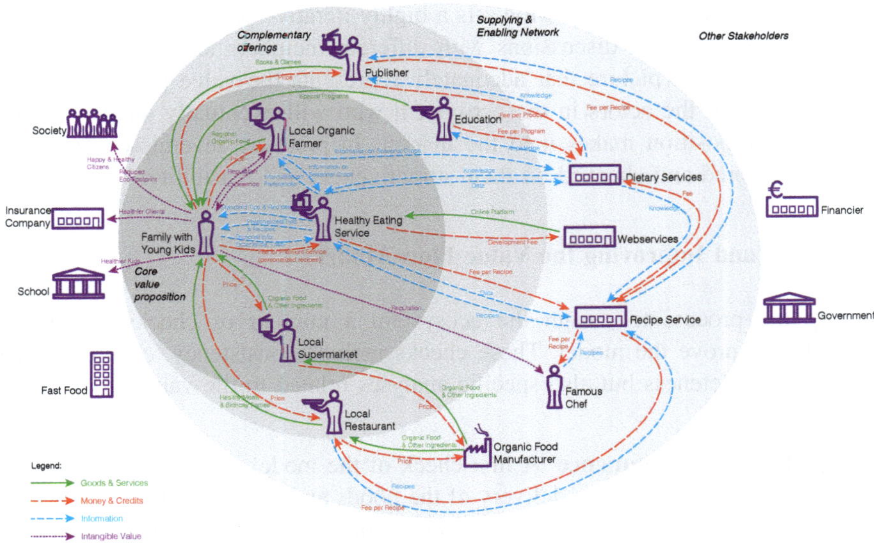

Fig. 6.16 Value Flow Model with the transactions of value between the actors

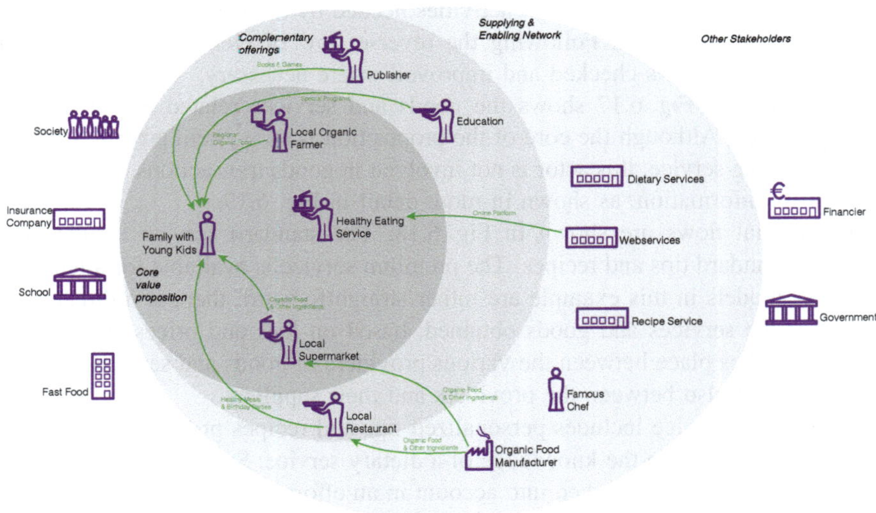

Fig. 6.17 Example of a check on completeness of goods and services flow

whether the required ingredients are available from a particular farmer or local supermarket at that specific time, or whether it is sufficient to just use crops that are generally available in that season (without the actual availability in specific shops). This of course has an impact on the information flows, and on the required activities by the actors to provide the information.

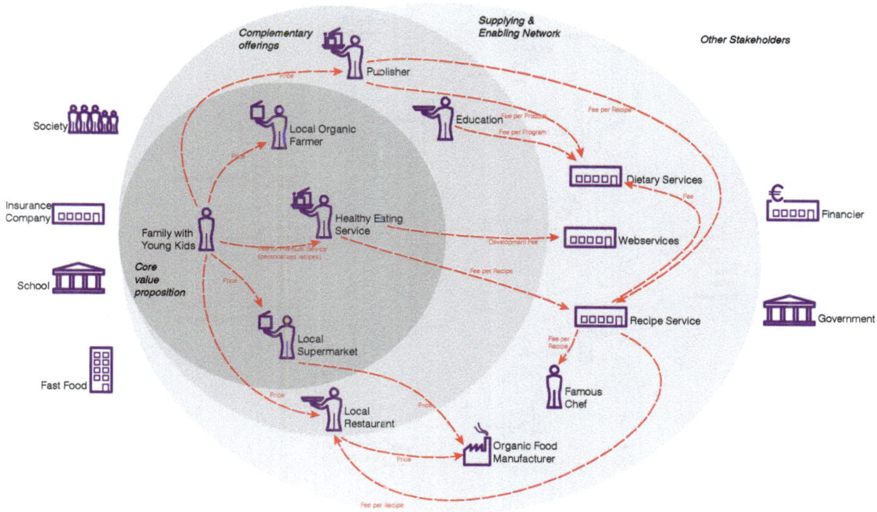

Fig. 6.18 Check on completeness of financial flow

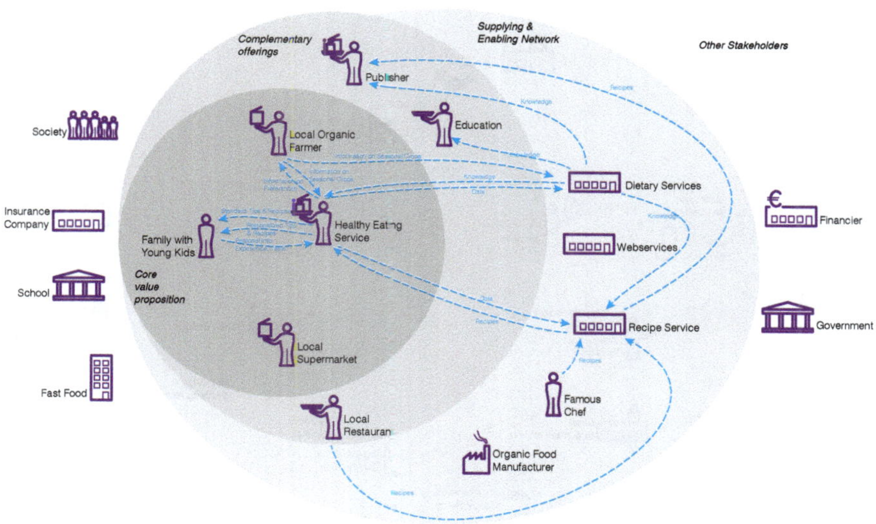

Fig. 6.19 Check on completeness of information flow

Figure 6.20 shows intangible flows, which also involve other stakeholders than those involved in the direct transactions in the value proposition.

To indicate how different decisions lead to a different Value Flow Model, the same value proposition is shown again in Fig. 6.21, but this time with the more controlled interface towards the customer, as used earlier in Fig. 6.10. In this scenario,

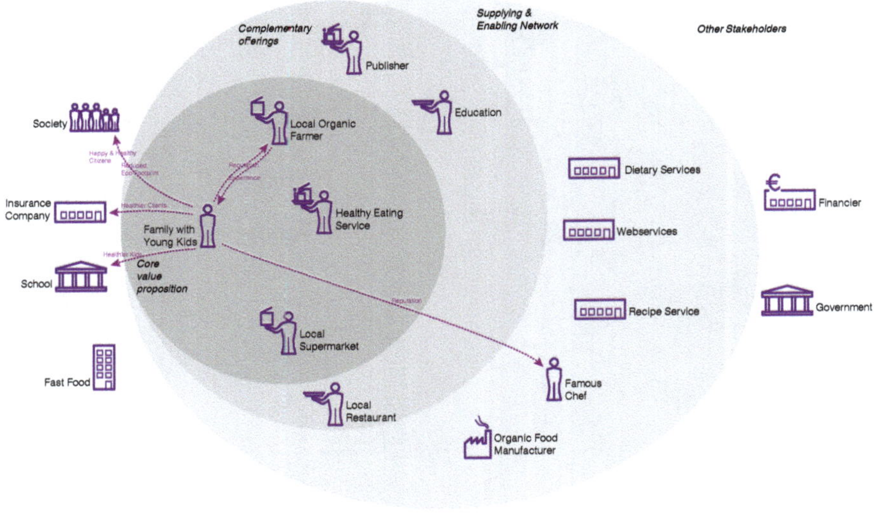

Fig. 6.20 Check on completeness of intangible flow

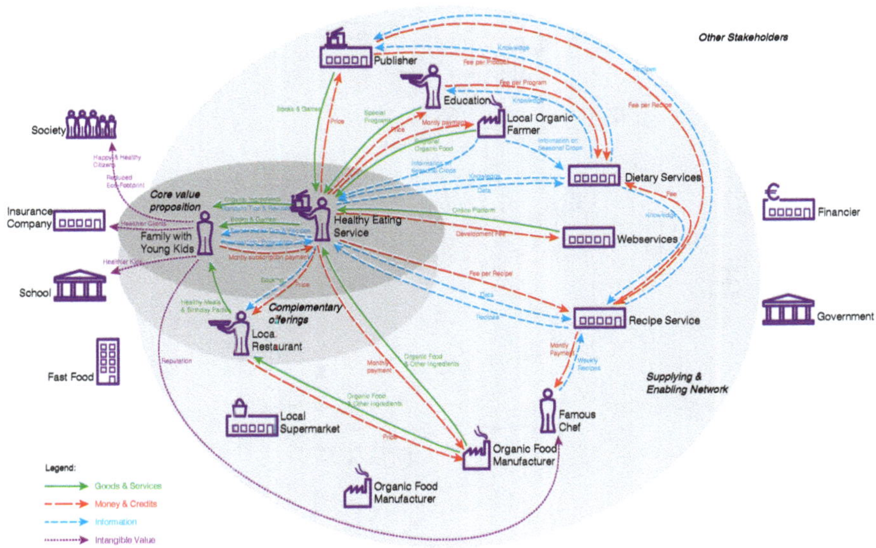

Fig. 6.21 Alternative scenario for value flows for the same value proposition as in Fig. 6.16

the customer is only dealing with the healthy eating service that coordinates the complete delivery of advice, recipes, books, games, regionally grown organic crops and other organic ingredients. In this scenario the customer pays a monthly subscription fee for the service, which covers the total costs of the package. The

Fig. 6.22 Check on reciprocity, using the example of a local organic farmer

systems provider has different business models with various actors: some also based on monthly payments, others on direct exchange. The risks and investments in this scenario for the systems provider will be higher than in the previous scenario. On the other hand the customer has to deal with only one provider, which makes it easier to create a coherent experience and offers more convenient shopping. The customer might also regard the experience of visiting a local organic farmer with his children to buy the ingredients as valuable. A selection of one scenario or the other will therefore require not only a check on the attractiveness not only from the business perspective, but surely also from that of the customer. Both scenarios could also be tested in the field with actual customers, and one customer group could be found to prefer one scenario and another group the other. The team can then decide whether it makes sense to offer both in parallel.

Reciprocity A second check involves the balancing of value in the ecosystem. Here, the aim is to make sure that there is a sound business model underlying the ecosystem. Each actor who invests value should receive a stream of value in return. This activity means that a short investigation is made of the value that each actor provides to the system (outgoing transaction arrows) and the value that he gains from the system (ingoing transaction arrows) (Fig. 6.22). Of course these arrows can be of a completely different nature: goods, money, information and intangible value can all be part of the transactions. Hence the check on the balance is not just a mathematical exercise. The best way to check the balance is 'if it feels good'. Intuition and gut feeling are better judgment criteria in this respect than trying to put a monetary value on intangible value. The perceived balance needs to fit with the motivations and values of the actors. For example, somebody for whom living a meaningful life is important, and whose goal it is to increase his well-being, may be less interested in gaining money from the transactions than somebody who aims only for wealth.

The sustainability of the ecosystem in the longer term depends on the health of its members. The reciprocity check is therefore an important analysis in assessing the overall health of the system.

Transaction Governance Transaction governance refers to the way the flows of goods, resources or information are controlled by the relevant parties. It involves a discussion on which of the business actors will be directly facing the customers, and how the coherence in experience for the users and customers is arranged. In such an analysis it may be decided that some actors will not be directly facing the customers, but will offer their part of the value proposition through another actor. An example of this arrangement is Apple's iTunes, where developers of apps do not sell their app directly to the users, but through Apple's App Store on iTunes. This makes it possible for Apple to check the functional quality of the app, and to check the coherence in user interface and graphic qualities. In this role Apple acts as a gatekeeper to ensure the quality of the total value proposition.

Scalability An important topic to consider for the design of the ecosystem is its ability to scale up when the market develops. Going through the ecosystem systematically to identify bottlenecks and the level of volume at which they appear is a useful exercise, that also highlights some risks related to the sustainability of the total system. Such an analysis usually takes place by going through a number of scenarios. Typical questions may include: What volumes are needed to recover the required investments?, and How fast can the different actors realize these volumes? What if the market is picking up much faster than anticipated—where are the bottlenecks, and what could be the solutions? And what if the market is adopting the innovation slower than anticipated—which actors will experience problems and what could be done to overcome such problems?

Flexibility is also an important aspect: What differences in volumes can the total system handle? What if there is a sudden peak in new customers wanting to purchase the proposition: can those new customers be accommodated? What if there is a slow period? Where are important dependencies in the system?

Business Models at Actor Level

Related to the financial flows in the ecosystem is the discussion of the business models at actor level. In fact a business model should be selected for each actor in the system that fits with the role. It is not necessary that the overall business model of the value proposition as presented to the customers is the same as the detailed business models between the actors. For example the overall proposition could be based on subscriptions, with customers paying a fixed monthly fee to the system provider. But the system provider can obtain the hardware for the system from a device manufacturer, and pay this supplier per product.

To a certain extent the business models at detailed level can be designed in this step, but it is also possible that they depend strongly on the particular organization playing the actor's role. As the following section shows, particular organizations may also combine certain roles, and this may also influence the business model.

With the indication of the actors and the transactions and business models at actor level, a first rather complete picture of the total ecosystem and its value network is emerging. This picture is still independent of the specific organizations playing the roles. At this stage a final check can be made of whether all the value elements are sufficiently covered for all the levels of the innovation, by critically assessing the result with the help of the Value Framework (Fig. 6.23). This may indicate some areas that are insufficiently covered, or have been insufficiently checked with the stakeholders. This step also includes checking once again the

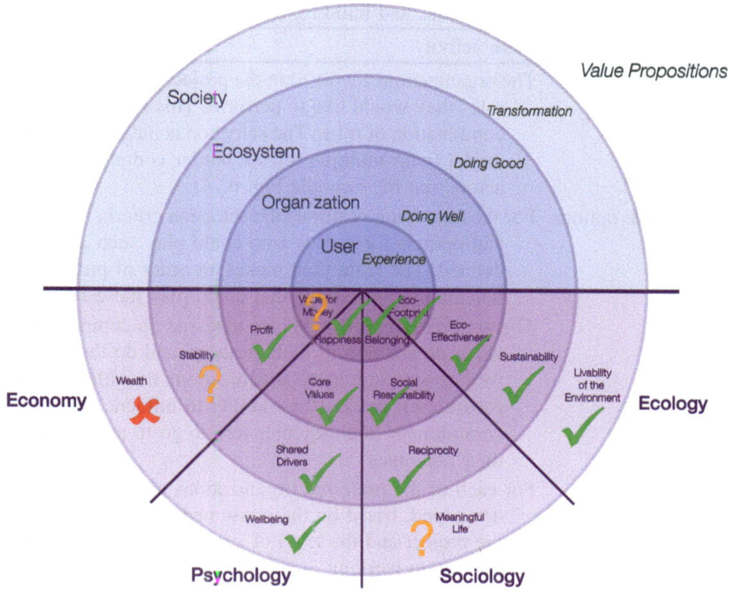

Fig. 6.23 Checking the value elements of the healthy eating service with the Value Framework

motivations of the members of the project team to contribute to the value proposition and their aspirations in the longer term, to ensure alignment towards a common goal and the existence of shared drivers. Once the picture is clear and complete it is time to start dividing the cake.

Selecting Parties for Realization and Implementation

The final step towards realizing the ecosystem is the selection of parties for the realization and implementation. So far the investments have been limited, but when making the shift towards the actual realization, the development of the goods and services will need to be executed. Likewise, production facilities, infrastructures and distribution channels will need to be set up, as is the case in any innovation project. A Lot of literature is available on how to execute such projects, also in collaborative settings. Hence in this section the focus will be on the specific activities related to the ecosystem development.

Selecting Parties

It is only natural that the team members who have been very active in bringing the project to this stage have the first choice of the activities in the total ecosystem. As they are usually very aware of the interdependency of all parties in the ecosystem, they will more or less explicitly address the capabilities of each of the

Table 6.2 Main steps in selecting parties and building the ecosystem

Step	Main activity
First pick	The organizations involved in the project so far indicate which actor roles they would like to perform. This may be a single role or a combination of roles. The selection is indicated in the value flow model (with multiple roles a border is drawn around those actors, see for example Fig. 6.24)
Selection criteria & options	For the remaining positions the selection criteria need to be defined, and options for parties who could play such a role should be defined. The team then makes an order of preference for the potential organizations that could play these roles
Screening	The preferred organizations are assessed on compatibility (does the value proposition fit their strategy, and do the values and the way of working of the organizations fit with the emerging ecosystem); power (their ability to influence the project); and commitment (are they prepared to go to any lengths to achieve their outcomes)
Develop engagement strategy	For each of the preferred organizations an engagement strategy is developed, based on their assumed interest in the outcomes of the project and the level of support that is needed to get the ecosystem working
Approaching the targeted organizations	Starting with the key positions in the network, the team approaches the targeted organizations, either directly or through an intermediary, and discusses the attractiveness for the organization to participate in the ecosystem, and where possible makes agreements with the selected parties
Building commitment	More information is exchanged, aspirations are shared, relationships established and expectations of both sides are clarified
Adapting the Value Flow Model	The organizations now entering the ecosystem may have an influence on the activities and transactions, so the Value Flow Model is updated with these insights, and a check is made of its consistency and attractiveness
Balancing value for all parties	As a final step, a check is made of the balance between the value each party has to provide to the system and the value it gains from the system

organizations choosing to play a role in the implementation and realization. The process usually follows a sequence of steps [14, 21, 62, 78], as shown in Table 6.2.

In the example of the healthy eating service, a feasible combination for one of the parties would be to combine the roles of system provider with those of the dietary and recipe service, as shown in Fig. 6.24.

Balancing Value

When all positions have been filled and all parties are known, it is important to check whether there is a sound balance of value in the ecosystem: do all the parties perceive the total system to be fair, and is there a sufficient level of reciprocity (see

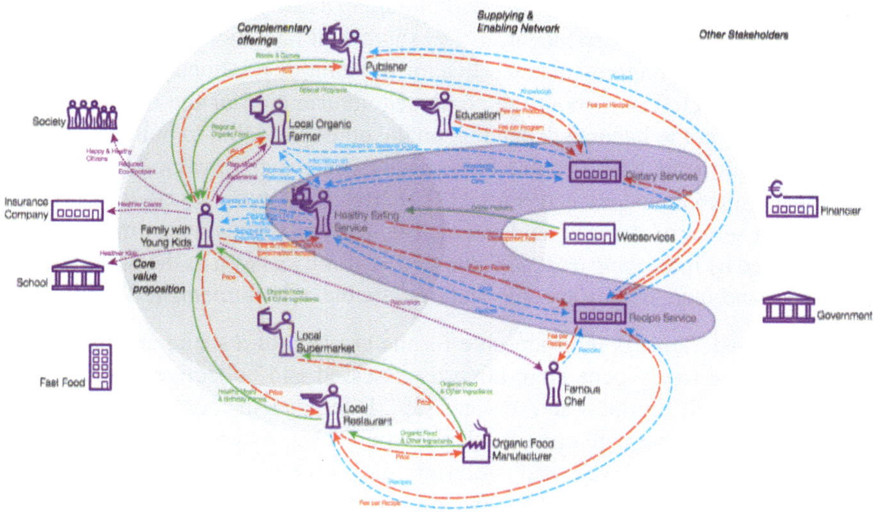

Fig. 6.24 Combining roles in the example of the healthy eating services

also the specific section on this subject). All actors have to be better off in the new ecosystem compared with the existing value creation system. The higher the number of actors, the more difficult it becomes to distribute the value created and to manage the value-creating constellation. Moreover, innovations may require huge adjustments by a number of key players. The larger the adjustments and investments required for the commercialization of the innovation, the tougher the management challenges will be. It may even be necessary to compensate companies that are worse off in the new business ecosystem [95].

A recurring problem during the initial phase of a new ecosystem is a thin market: buyers (providers of goods or services who buy components from suppliers) may be discouraged by an erratic or insufficient supply, while the suppliers face a market that is too thin to support large enough premiums (when they risk having excess supply). In that case ensuring pull-through demand to mitigate the suppliers risk may be a convenient way to get through this initial phase [95].

Specific arrangements may be needed in the ecosystem to ensure that all parties have a sustainable position both at the start-up of the ecosystem and in the longer term, and that they can contribute to making the total ecosystem flourish. These arrangements should be included in the business models between the actors and the total value flows in the system.

Building Relationships

It is quite likely that both small and large organizations are involved in the ecosystem. Relationships between large and small organizations are traditionally transaction-based: they can be terminated at the will of either (usually the larger) party. But in ecosystems both are interdependent, so they share their fate, each party relying on the other in a sustained, ongoing manner. In a truly symbiotic relationship, neither party can perform effectively without the continued contribution of the other [35]. The critical issue is not to become dependent on the other party, but to build and sustain mutual commitments which enable both partners to engage in value-creating coordination of interdependent activities [12]. Partners in more successful alliances engage in a series of iterative and interactive learning cycles over time, typically characterized by increasing trust and adaptive flexibility as well as the willingness to take larger and increasingly specific and irreversible commitments [33].

Building Trust Trust is indispensable for the creation of a social environment in which ideas are freely generated, honestly assessed and collectively transformed into profitable new products and services [30]. Dilemmas in ecosystems arise from the risk of cooperating with other parties. Feelings people experience are related to the fear of being betrayed and suffering the consequences, the attraction of achieving a mutual goal together, and probably also a temptation to betray the other for personal gain. The process of building trust to overcome potential difficulties in the project starts at the very beginning. If all parties are involved in defining the problem situation for which the project aims to find a solution, and the solution is being defined using the intelligence and experience of all parties, trust will start to build between them right from the start. One way to support the process of building trust is to put the sentiments of all the parties into words by answering three questions: what are your fears, your attractions and your temptations? [63].

An important foundation of trust is respect for the other. This implies viewing others irrespective of their status as equals, and of deserving their needs. It requires transcending self-preoccupation [30]. Trust cannot be mandated; it needs to be earned and develops gradually. It is more likely to develop in contexts where the parties demonstrate their capability to fulfill a promise, are frank in interactions, live up to expectations and are honest with each other [17].

Problem-Solving Problems will inherently be phased when a team embarks on a process of developing radical innovations and their ecosystem. One common issue is that the members participating in the project represent their own organizations, and do not always have the full support of their organizations in making decisions. A practice that has been very successful in the industry when working with collaborative innovation between people of different cultures is to dedicate a workshop to defining the values and behavior of the project team in such situations. The workshop results in a short document in which the team describes behavior, roles, tasks and responsibilities in case of:

- Conflicts within the team
- Conflicts between members of the team and people outside the team
- Agreements with respect to necessary commitments of the team
- Internal communication

- External communication
- Risk handling and sharing in case of major changes (project ambition, organization, planning etc.).

Holding such a discussion before problems arise helps to build the commitment of the members. It also helps to define what to do if the team members face pressure from their own organization. If these potential problems have been discussed in advance, members will find it less difficult to raise issues when and if they happen.

If conflicts arise it is important not just to rely on contracts, but rather try to find solutions that meet the needs of the parties involved in a win–win spirit. Handling unforeseen problems and conflicts in a way that shows respectability for the well-being of the ecosystem is vital [17].

Evolution of Ecosystems

Ecosystems are dynamic in nature; they continuously evolve, adapting to new conditions that arise from changes in the environment [48]. The environment may change in ways that make it genuinely less attractive to at least one of the partners [31]. One of the members of the ecosystem may then decide to exit, and the ecosystem has to adapt to the new situation.

James Moore [69] has given an interesting description of the evolution of ecosystems in four distinct stages: birth, expansion, leadership and self-renewal or death. As this view is largely defined from a company perspective, it has been tuned towards ecosystems for meaningful innovations.

Stage 1: Birth In this stage an individual innovator or group of people starts to define a vision for a new product or service that aims to solve a societal problem. The focus is on what customers want, in other words the value of a proposed new product or service and the best form in which to deliver it. It often pays to cooperate in this stage: business partners help fill out the full package of value for customers. And by attracting important 'follower' companies, leaders may stop them from helping other emerging ecosystems. But co-evolving companies must do more than satisfy customers; a leader must also emerge to initiate a process of rapid, ongoing improvement that draws the entire community toward a grander future. Established companies are often better off waiting and watching carefully as a new market sorts itself out. The iterative process of trying out innovative ideas and discovering which solutions are attractive to customers is hard to accomplish in traditional corporate culture. Established companies can subsequently replicate successful ideas and broadcast them across a wider market. In other words, they can enter the market at stage 2 by appropriating the developmental work of others. Cooperative challenges: working with customers and suppliers to define the new value proposition around a seed innovation. Competitive challenges: protecting your ideas from others who might be working towards defining similar offers; and tying up critical lead customers, key suppliers and important channels.

Stage 2: Expansion Business ecosystems expand to conquer broad new territories. For meaningful innovations this is an important stage, as it increases the impact at societal level: the more the adopters of the innovation, and the more organizations becoming active members of the ecosystem, the higher the impact. Some business expansions meet little resistance, but in other cases rival ecosystems may be closely matched and choose to attack the same territory. Direct battles for market share break out. In the end one business ecosystem may triumph, or rival ecosystems may reach semi-stable accommodations. In general, two conditions are necessary for expansion: a business concept that a large number of customers will value; and the potential to scale up the concept to reach this broad market. During the expansion stage, established companies can exercise enormous power in marketing and sales, as well as in the management of large-scale production and distribution, literally crushing smaller ecosystems in the process. One of the most important managerial challenges is to stimulate market demand without greatly exceeding the ability to meet it. In general, stage 2 rewards fast expansion that squeezes competing ecosystems into the margins. But managers must also prepare for future leadership and leverage in the next stage. To do so, companies must maintain full control of customers' relationships and core centers of value and innovation. Moreover, they must develop relationships with their suppliers that constrain these followers from becoming leaders in stage 3. Cooperative challenges: bringing the new offer to a large market by working with suppliers and partners to scale up supply and to achieve maximum market coverage. Competitive challenges: defeating alternative implementations of similar ideas; ensuring that your approach is the market standard in its class through dominating key market segments.

Stage 3: Leadership Two conditions contribute to the onset of the leadership struggles that are the hallmark of stage 3: the ecosystem must have strong enough growth and profitability to be considered worth fighting over; and the structure of the value-adding components and processes that are central to the business ecosystem must become reasonably stable. This stability allows suppliers to target particular elements of value and to compete in contributing them. It encourages members of the system to consider expanding by taking over activities from those closest to them in the value chain. Most of all, it reduces the dependence of the whole ecosystem on its original leader. It is in stage 3 that companies become preoccupied with standards, interfaces, 'the modular organization', and customer–supplier relations. Bargaining power comes from having something the ecosystem needs, and being the only practical source of it. Sometimes this sole-source status can be established contractually or through patent protection. But fundamentally, it depends on constant innovation—on creating value that is critical to the whole ecosystem's continued price/performance improvement. Central ecological contributors maintain the much-coveted stranglehold within a business ecosystem. In brief, other members cannot live without them. This central position enables them to bargain for a higher share of the total value produced by the ecosystem. Central contributor status is maintained in part by the investments others have made in being followers. Central companies reinforce their roles by making important innovative contributions to the performance of the ecosystem as whole. Finally, followers value a central contributor because of its grip on customers. In any case, for dominant companies the expansion and leadership stages of an ecosystem can make or break them. In stage 3, lead producers must extend control by continuing to shape future directions and the investments of key customers and suppliers. And for healthy profits, any company in the ecosystem—leader or follower—must maintain bargaining power over other members. Cooperative challenges: providing a compelling vision for the future that encourages suppliers and customers to work together to continue improving the complete offer. Competitive challenges: maintaining strong bargaining power in relation to other players in the ecosystem, including key customers and valued suppliers.

Stage 4: Self-Renewal (or Death) Mature business ecosystems are threatened by rising new ecosystems and innovations. Alternatively, a community might undergo the equivalent of an earthquake: sudden new environmental conditions that include changes from governmental regulations, customer buying patterns or macroeconomic conditions. Moreover, these two factors reinforce each other. A changed environment is often more hospitable to new or formerly marginal business ecosystems. In fact, how a dominant company deals with the threat of obsolescence is the ultimate challenge. Leading successive generations of innovations is clearly crucial to an ecosystem's long-term success and its ability to renew itself. There are three general approaches to self-renewal, which can be used alone or in combination: dominant companies can seek to slow the growth of a new ecosystem; they can try to incorporate new innovations into their own ecosystems; or they can fundamentally restructure themselves to try coping with a new reality. Of course, the leaders of a mature business ecosystem sometimes have no choice but to undertake profound structural and cultural changes. Despite the difficulties of a complex business environment, managers can design longevity into an ecosystem. During the expansion and leadership stages, for example, companies can work hard to micro-segment their markets, creating close, supportive ties with customers. These customers will remain committed to a particular ecosystem long enough for its members to incorporate the benefits of new approaches. Cooperative challenges: working with innovators to bring new ideas to the existing ecosystem. Competitive challenges: maintaining high barriers to entry to prevent other innovators from building alternative ecosystems; maintaining high customer switching costs to buy time to incorporate new ideas into your own products and services.

Although it might sound challenging to embark on the journey of designing value propositions that bring value to people, organizations and society, it is certainly possible to do so. There are big gains to be made in the process: it is much more rewarding to work on solutions that actually improve the lives of people and create business at the same time, than to work on just another product for profit alone.

This book aims to provide a solid background on value creation and the process of designing new value propositions, business models and ecosystems for meaningful innovations. It provides practical tools that support people in the process. But most of all it aims to inspire people.

With the help of creative and motivated people, supported by information and social technologies, and above all with the spirit and drive of visionary people, we can make this world a better place. Let us focus our energy on realizing what Aristotle defined as: "the art of living, and living well".

References

1. Adner, Ron. 2006. Match your innovation strategy to your innovation ecosystem. *Harvard Business Review*, April 2006.
2. Adner, Ron, and Rahul Kapoor. 2010. Value creation in innovation ecosystems: How the structure of technological interdependence affects firm performance in new technology generations. *Strategic Management Journal* 31(3): 306–333.
3. Allee, Verna. 2008. Value network analysis and value conversion of tangible and intangible assets. *Journal of Intellectual Capital* 9(1): 5–24.
4. Allee, Verna. 2000. Reconfiguring the value network. *Journal of Business Strategy* 21(4): 36–39. July–Aug 2000.
5. Amit, Raphael, and Christoph Zott. 2001. Value creation in e-business. *Strategic Management Journal* 22: 493–520.
6. Andersson, Birger, Maria Bergholtz, and Paul Johannesson. 2009. Purpose driven value model design. In: Proceedings of the CAiSE workshop BUSITAL.
7. Bäckstrand, K. 2006. Multi-stakeholder partnerships for sustainable development: Rethinking legitimacy, accountability and effectiveness. *European Environment* 16(5): 290–306.
8. Ballantyne, David, and Richard J. Varey. 2006. Creating value-in-use through marketing interaction: The exchange logic of relating, communicating and knowing. *Marketing Theory* 6(3): 335–348.
9. Ballon, Pieter. 2007. Business modelling revisited: The configuration of control and value. *Info* 9(5): 6–19.
10. Bartel, Caroline A., and Raghu Garud. 2009. The role of narratives in sustaining organizational innovation. *Organization Science* 20(1): 107–117.
11. Basole, R.C., and W.B. Rouse. 2008. Complexity of service value networks: Conceptualization and empirical investigation. *IBM Systems Journal* 47(1): 53–70.
12. Blankenburg-Holm, Desiree, Kent Eriksson, and Jan Johanson. 1999. Creating value through mutual commitments to business network relationships. *Strategic Management Journal* 20: 467–486.
13. Blau, Benjamin, Christof Weinhart, Clemens van Dinther, Tobias Conte, and Yongchun Xu. 2009. How to coordinate value generation in service networks— a mechanism design approach. *Business & Information Systems Engineering* 5: 343–356.
14. Bourne, Lynda, and Derek H.T. Walker. 2006. Using a visualisation tool to study stakeholder influence—two Australian examples. *Project Management Journal* 37(1): 5–21.
15. Brandenburger, Adam M., and Barry J. Nalebuff. 1997. *Co-opetition: 1. A revolutionary mindset that redefines competition and cooperation; 2. The game theory strategy that's changing the game of business*. New York: Crown Business Publishers.
16. Brugha, Ruairi, and Zsuzsa Varvasovszky. 2000. Stakeholder analysis: A review. *Health Policy and Planning* 15(3): 239–246.

17. Bstieler, Ludwig. 2006. Trust formation in collaborative new product development. *Journal of Product Innovation Management* 23: 56–72.
18. Bunn, Michele D., Grant T. Savage, and Betsy B. Holloway. 2002. Stakeholder analysis for multi-sector innovations. *The Journal of Business & Industrial Marketing* 17(2/3): 181–203.
19. Chesbrough, Henry. 2006. *Open business models: How to thrive in the new innovation landscape*. Boston: Harvard Business Press.
20. Clarkson, Max B.E. 1995. A stakeholder framework for analyzing and evaluating corporate social performance. *Academy of Management Review* 20(1): 92–117.
21. Cova, Bernard, and Robert Salle. 2008. Marketing solutions in accordance with the S-D logic: Co-creating value with customer network actors. *Industrial Marketing Management* 37(3): 270–277.
22. Cravens, David W., Nigel F. Piercy, and Shannon H. Shipp. 1996. New organizational forms for competing in highly dynamic environments: The network paradigm. *British Journal of Management* 7: 203–218.
23. Dahlander, Linus, and David M. Gann. 2010. How open is innovation? *Research Policy* 39: 699–709.
24. De Boever, Jorn. Value networks of P2P TV: An analysis of actors and their roles. 3rd International Conference on Internet and Web Applications and Services, June 8–13, 2008.
25. De Reuver, Mark, Harry Bouwman, and Timber Haaker. Capturing value from mobile business models: Design issues that matter. 21st Bled eConference eCollaboration: Overcoming Boundaries through Multi-Channel Interaction, Bled, Slovenia, June 15–18, 2008.
26. De Rond, Mark, and Hamid Bouchikhi. 2004. On the dialectics of strategic alliances. *Organization Science* 15(1): 56–69.
27. Den Ouden, Elke, and Rianne Valkenburg. Value models in social open innovation. Proceedings of the 21st ISPIM Conference, Bilbao, Spain, June 6–9, 2010.
28. Doganova, Liliana, and Marie Eyquem-Renault. 2009. What do business models do? Innovation devices in technology entrepreneurship. *Research Policy* 38: 1559–1570.
29. Donaldson, Thomas, and Lee E. Preston. 1995. The stakeholder theory of the corporation: Concepts, evidence and implications. *Academy of Management Review* 20(1): 65–91.
30. Dovey, Ken. 2009. The role of trust in innovation. *The Learning Organization* 16(4): 311–325.
31. Doz, Yves L. 1996. The evolution of cooperation in strategic alliances: Initial conditions or learning processes? *Strategic Management Journal* 17: 55–83. Special Issue: Evolutionary Perspectives on Strategy, Summer.
32. Doz, Yves L., and Mikko Kosonen. 2010. Embedding strategic agility: A leadership agenda for accelerating business model renewal. *Long Range Planning* 43: 370–382.
33. Doz, Yves, Jose Santos, and Peter Williamson. 2001. *From global to metanational: How companies win in the knowledge economy*. Boston: Harvard Business Press.
34. Emden, Zeynep, Roger J. Calantone, and Cornelia Droge. 2006. Collaborating for new product development: Selecting the partner with maximum potential to create value. *Journal of Product Innovation Management* 23: 330–341.
35. Etemad, Hamid, Richard W. Wright, and Leo Paul Dana. 2001. Symbiotic international business networks: Collaboration between small and large firms. *Thunderbird International Business Review* 43(4): 481–499.
36. Florida, Richard. 2002. *The rise of the creative class:… and how it's transforming work, leisure, community, and everyday life*. New York: Persues Books.
37. Freel, Mark, and Jeroen P.J De Jong. 2009. Market novelty, competence-seeking and innovation networking. *Technovation* 29: 873–884.
38. Freeman, R.Edward. 1984. *Strategic management: A stakeholder approach*. Boston: Pitman.
39. Freeman, R.Edward, Andrew C. Wicks, and Bidhan Parmar. 2004. Stakeholder theory and "the corporate objective revisited". *Organization Science* 15(3): 364–369.
40. Freeman, R.Edward, Jeffrey S. Harrison, and Andrew C. Wicks. 2010. *Stakeholder theory. The state of the art*. Cambridge: Cambridge University Press.

41. Geels, Frank W. 2004. From sectoral systems of innovation to socio-technical systems. Insights about dynamics and change from sociology and institutional theory. *Research Policy* 33: 897–920.
42. Gloor, Peter A. 2006. *Swarm creativity: Competitive advantage through collaborative innovation networks*. New York: Oxford University Press.
43. Gordijn, Jaap, and Hans Akkermans. 2001. Designing and evaluating e-business models. *IEEE Intelligent Systems* 16: 11–17.
44. Gordijn, Jaap, Hans Akkermans, and Hans van Vliet. 2000. *Business modelling is not process modelling. In conceptual modeling for e-business and the web, lecture notes in computer science*, 40–51. Berlin/Heidelberg: Springer.
45. Grumbine, R.Edward. 1994. What is ecosystem management? *Conservation Biology* 8(1): 27–38.
46. Haaker, Timber, Edward Faber, and Harry Bouwman. 2006. Balancing customer and network values in business models for mobile services. *International Journal of Mobile Communications* 4(6): 645–661.
47. Hakansson, Hakan, and David Ford. 2002. How should companies interact in business networks? *Journal of Business Research* 55(2): 133–139.
48. Hakansson, Hakan, David Ford, Lars-Erik Gadde, Ivan Snehota, and Alexandra Waluszewski. 2009. *Business in networks*. London: Wiley.
49. Haksever, Cengiz, Radha Chaganti, and Ronald G. Cook. 2004. A model of value creation: Strategic view. *Journal of Business Ethics* 49: 291–305.
50. Haldane, Andrew G., and Robert M. May. 2011. Systemic risk in banking ecosystems. *Nature* 469: 351–355.
51. Hansen, John D., and Michele D. Bunn. 2009. Stakeholder relationship management in multi-sector innovations. *Journal of Relationship Marketing* 8: 196–217.
52. Hansen, Morten T., Henry W. Chesborough, Nitin Nohria, and Donald N. Sull. 2000. Networked incubators. Hothouses of the new economy. *Harvard Business Review* 78: 74–84.
53. Harrison, Jeffrey S., and R.Edward Freeman. 1999. Stakeholders, social responsibility, and performance: Empirical evidence and theoretical perspectives. *Academy of Management Journal* 42(5): 479–485.
54. Hearn, Greg, and Cassandra Pace. 2006. Value-creating ecologies: Understanding next generation business systems. *Foresight* 8(1): 55–65.
55. Iansiti, Marco, and Roy Levien. 2004. *The keystone advantage: What the new dynamics of business ecosystems mean for strategy, innovation, and sustainability*. Boston: Harvard Business Press.
56. Iansiti, Marco, and Roy Levien. 2004. Strategy as ecology. *Harvard Business Review* 82(3): 68–78.
57. Ihlstrom Eriksson, C., M. Akesson, M. Bergquist, and J. Ljungberg. Forming a value network-analyzing the negotiation between actors in the e-newspaper case. Proceedings of the 42nd Hawai'i International Conference on System Sciences HICSS-42, Big Island, HI, USA, January 5–8, 2009.
58. Johnson, Mark W., Clayton M. Christensen, and Henning Kagermann. 2008. Reinventing your business model. *Harvard Business Review* 86: 50–59.
59. Jonas, W., Morelli, N. and Munch, J. Designing a product service system in a social framework - methodological and ethical considerations. Proceedings of Undisciplined! Design Research Society Conference, Sheffield Hallam University, United Kingdom, July 2008.
60. Kao, John. 2009. Tapping into the world's innovation hotspots. *Harvard Business Review* 87: 109–114.
61. Kroes, Peter, Maarten Franssen, Ibo van de Poel, and Maarten Ottens. 2006. Treating socio-technical systems as engineering systems: Some conceptual problems. *Systems Research and Behavioral Science* 23(6): 803–814.
62. Krucken, Lisa, and Anna Meroni. 2006. Building stakeholder networks to develop and deliver product-service-systems: Practical experiences on elaborating pro-active materials for communication. *Journal of Cleaner Production* 14: 1502–1508.

63. Le Cardinal, Gilles, Jean-Francois Guyonnet, Bruno Poauzoullic, and Julie Rigby. 2001. Intervention methodology for complex problems: The FAcT-mirror method. *European Journal of Operational Research* 132: 694–702.
64. Ledwith, Ann, and Paul Coughlan. 2005. Splendid isolation: Does networking really increase new product success? *Creativity and Innovation Management* 14(4): 366–373.
65. Lee, Sungjoo, Gwangman Park, Byungun Yoon, and Jinwoo Park. 2010. Open innovation in SMEs: An intermediated network model. *Research Policy* 39: 290–300.
66. Li, Feng, and Jason Whalley. 2002. Deconstruction of the telecommunications industry: From value chains to value networks. *Telecommunications Policy* 26(9–10): 451–472.
67. Logan, Robert K., and Greg Van Alstyne. 2007. Design ecology: designing for emergence and innovation II. Working paper.
68. Moore, James F. 1993. Predators and prey: A new ecology of competition. *Harvard Business Review* 71: 75–86.
69. Moore, James F. 1996. *The death of competition: Leadership and strategy in the age of business ecosystems*. New York: HarperCollins Publishers.
70. Morelli, Nicola. 2007. Social innovation and new industrial contexts: Can designers "industrialize" socially responsible solutions? *Design Issues* 23(4): 3–21. Autumn 2007.
71. Nambisan, Satish, and Mohanbir Sawhney. 2008. *The global brain: Your roadmap for innovating faster and smarter in a networked world*. NJ: Pearson Prentice Hall.
72. Osterwalder, Alexander, and Yves Pigneur. 2010. *Business model generation: A handbook for visionairies, game changers, and challengers*. New York: Wiley.
73. Ottens, Maarten, Maarten Franssen, Peter Kroes, and Ibo Van De Poel. 2006. Modelling infrastructures as socio-technical systems. *International Journal of Critical Infrastructures* 2(2–3): 133–145.
74. Oxley Green, Abigail, and Lynsey Hunton-Clarke. 2003. A typology of stakeholder participation for company environmental decision-making. *Business Strategy and the Environment* 12: 292–299.
75. Parameswaran, Lekshmy, and Jeroen Raijmakers. 2010. People-focused innovation in healthcare. How philips design supports development of solutions for the ever-changing healthcare landscape. Philips Design. http://www.design.philips.com/philips/shared/assets/design_assets/pdf/nvbD/july2010/people_focused_innovation_in_healthcare.pdf.
76. Payne, Adrian, David Ballantyne, and Martin Christopher. 2005. A stakeholder approach to relationship marketing strategy: The development and use of the "six markets" model. *European Journal of Marketing* 39(7/8): 855–871.
77. Peltoniemi, Mirva, and Elisa Vuori. 2004. Business ecosystem as the new approach to complex adaptive business environments. Proceedings of the eBusiness Research Forum, 267–281.
78. Peppard, Joe, and Anna Rylander. 2006. From value chain to value network: Insights for mobile operators. *European Management Journal* 24(2–3): 128–141. April-June 2006.
79. Pickett, S.T.A., and M.L. Cadenasso. 2002. The ecosystem as a multidimensional concept: Meaning, model, and metaphor. *Ecosystems* 5: 1–10.
80. Pisano, Gary P., and Roberto Verganti. 2008. Which collaboration is right for you? *Harvard Business Review* 86: 78–86.
81. Polonsky, Michael Jay, Les Carlson, and Marie-Louise Fry. 2003. The harm chain: A public policy development and stakeholder perspective. *Marketing Theory* 3(3): 345–364.
82. Poot, Tom, Dries Faems, and Wim Vanhaverbeke. 2009. Toward a dynamic perspective on open innovation: A longitudinal assessment of the adoption of internal and external innovation strategies in the Netherlands. *International Journal of Innovation Management* 13(2): 177–200.
83. Porter, Michael E., and Mark R. Kramer. 2011. Creating shared value. *Harvard Business Review* 89(1/2): 62–77. January.
84. Post, James E., Lee E. Preston, and Sybille Sachs. 2002. Managing the extended enterprise: The new stakeholder view. *California Management Review* 45(1): 6–28.

85. Powell, Walter W. 1990. Neither market nor hierarchy: Network forms of organization. *Research in Organizational Behavior* 12: 295–336.
86. Ramírez, R. 1999. Value co-production: Intellectual origins and implications for practice and research. *Strategic Management Journal* 20: 49–65.
87. Riedl, Christoph, Tilo Boehmann, Jan Marco Leimeister and Helmut Krcmar. 2009. A framework for analysising service ecosystem capabilities to innovate. 17th European Conference on Information Systems ECIS, Verona, Italy.
88. Ritala, Paavo, and Pia Hurmelinna-Laukkanen. 2009. What's in it for me? Creating and appropriating value in innovation-related coopetition. *Technovation* 29(12): 819–828.
89. Rowley, Timothy J. 1997. Moving beyond dyadic ties: a network theory of stakeholder influences. *Academy of Management Review* 22(4): 887–910.
90. Sawhney, Mohanbir, and Emanuela Prandelli. 2000. Communities of creation: Managing distributed innovation in turbulent markets. *California Management Review* 42(4): 24–54.
91. Stathel, Stephan, Jan Finzen, Christoph Riedl, and Norman May. 2008. Service innovation in business value networks. 18th International RESER Conference, Stuttgart.
92. Swink, Morgan. 2006. Building collaborative innovation capability. *Research Technology Management* 49(2): 37–47.
93. Tapscott, D., A. Lowy, and D. Ticoll. 2000. *Digital capital: Harnessing the power of business webs*. Boston: Harvard Business Press.
94. Thompson, James D., and Ian C. MacMillan. 2010. Business models: Creating new markets and societal wealth. *Long Range Planning* 43(2–3): 291–307.
95. Vanhaverbeke, W., and M Cloodt. 2007 Open innovation in value networks. Chapter in *Open innovation: Researching a new paradigm*, 258–285. Oxford: Oxford University Press.
96. Van Marrewijk, Marcel. 2004. A value based approach to organisation types: Towards a coherent set of stakeholder-oriented management tools. *Journal of Business Ethics* 55: 147–158.
97. Verganti, Roberto. 2008. Design, meanings and radical innovation: A meta-model and a research agenda. *Journal of Product Innovation Management* 25(5): 436–456.
98. Verstraete, Thierry, and Estele Jouison. 2006. Connecting stakeholders theory and conventions theory to highlight the adhesion of stakeholders to the business model of a start-up. International Council for Small Business, 51st World Conference, Melbourne, Australia, June 2006.
99. Von Bertalanffy, Ludwig. 1972. The history and status of general systems theory. *Academy of Management Journal* 15(4): 407–426.
100. Ward, Stephen, and Chris Chapman. 2008. Stakeholders and uncertainty management in projects. *Construction Management and Economics* 26: 563–577.
101. Weigand, Hans. 2009. Value Encounters–Modeling and Analyzing Co-Creation of Value. IFIP Conference e-Business, e-Services, and e-Society, Nancy, September 2009.
102. West, Joel. 2007. Value capture and value networks in open source vendor strategies. Proceedings of the Hawai'i International Conference on System Sciences HICSS-40, Waikola, Hawai'i, January 3–7, 2007.
103. Yunus, M., B. Moingeon, and L. Lehmann-Ortega. 2010. Building social business models: Lessons from the grameen experience. *Long Range Planning* 43: 308–325.
104. Zott, C., and R. Amit. 2010. Business model design: An activity system perspective. *Long Range Planning* 43: 216–226.

Index

A
Actor relationships, 174
Actors, 154
Adaptive system, 104
Adjacency strategies, 80
Agricultural economy, 6
Anthropology, 37
Aspiration, 110
Authenticity, 41

B
Behavior
  changing, 67
  consumer, 40
  model, 69
  sustainable, 71
  target, 68
Belonging, 43
Business model, 127, 180
  design parameters, 132
  innovation, 127
  profit formula, 136
  revenue model, 137
  scalable, 84
Buyers, 109

C
Challenges
  societal, 4
Change, 88
Commoditization trap, 26
Commodity, 26
Compatibility, 156
Complementary offerings, 159
Complementors, 152
Conflict, 184
Consumer
  behavior, 40
Control, 149
Core competences, 79
Core values, 36
Cost sharing, 174
Cradle-to-cradle design, 53
Creativity, 50
Culture, 39, 124
Customers, 167

D
Decisions
  rational, 25
Demonstrator, 106, 115, 117
Demonstrator planning, 106
Desirability, 104
Doing good, 85
Doing well, 81
Drivers
  shared, 36

E
Eco-effectiveness, 53
Eco-feedback, 71
Eco-footprint, 52
Ecology, 46
Economics, 23
Economy
  agricultural, 6
  experience, 7
  industrial, 6
  knowledge, 7
  transformation, 9

E. den Ouden, *Innovation Design*, DOI: 10.1007/978-1-4471-2268-5,

E (*cont.*)
Ecosystem, 65, 82, 143
  biology, 18, 145
  birth, 185
  building, 182
  building relationships, 183
  building trust, 184
  business, 17, 145
  death, 187
  decision making, 150
  dominator, 151
  evolution, 185
  expansion, 186
  industry, 144
  keystone, 151
  leadership, 151, 186
  management, 145
  niche player, 151
  paradoxes, 148
  project level, 144
  purpose, 143
  regional, 144
  roles, 171–172
  roles in the development team, 163
  self-renewal, 187
  stability, 152
  sustainability, 152
  types, 143
Emerging markets, 129
Empathy, 111
Enriching, 119, 162
Ethics, 73
Expeditionary marketing, 124
Experience, 62, 73
Experience economy, 7
Experience flow, 112
Experimentation, 115
Externalities, 28

F
Feasibility, 104
Financiers, 167
Forced functionality, 71

G
Gestalt theory, 101
Goods and services, 158
Greed, 42
Gross Domestic Product (GDP), 27–28
Gross National Happiness (GNH), 28, 51
Growth, 128

H
Happiness, 31, 35
Human development, 49

I
Ideation, 113
Implementation, 122
Industrial economy, 6
Influence, 156
Influencers, 110
Information, 72, 158
Innovation
  context aware, 72
  disruptive, 63
  end-user driven, 75
  market-driven, 147
  meaningful, 92, 97, 149
  open, 77, 144
  radical, 105
  social, 76, 87
Insight, 110
Intangibles, 159
Integrative thinking, 99
Interaction
  engaging, 71
  playful, 72
Interdependence, 152
Investments, 157

K
Knowledge economy, 7

L
Life satisfaction, 31
Livability of the environment, 54

M
Macroeconomics, 27
Market introduction, 122
Meaning, 41, 62, 89
Meaningful
  life, 46
Meaningful life, 34, 46
Meta-design, 103
Microeconomics, 24
Money, 25, 42
  credit, 158
Motivation, 155
Motivations, 155

N
Narcitropy, 42
Needs, 32, 49, 108
  fundamental, 24
  user, 24
Network, 146
  challenges, 148
  collaborative, 77
  company-centered, 146
  paradoxes, 148
  solution-driven, 146
  supplying and enabling, 161
  value, 77
New business creation, 126, 130

P
Paradigm shift, 5
Personas, 112
Persuasion, 68, 70
Philanthropy, 42
Planning, 106
Price, 62
Problem framing, 111
Problem-solving, 184
Profit, 30
Prototyping, 104
Psychology, 30
  positive, 34

R
Reciprocity, 41, 45, 159, 179
Reflective practice, 106
Responsibility
  social, 44
Revenue sharing, 174
Robustness, 85

S
Scalability, 180
Scripting, 71
Seduction, 70
Self-renewal, 78
Shared drivers, 36, 78
Shared value, 83, 90
Sketching, 102
Social responsibility, 44
Society at large, 167
Sociology, 37
Socio-technical service systems, 63
Solution, 62
  customer-facing, 63
  holistic, 101
Spirituality, 50
  business, 51
Stability, 30
Stage-gate process, 122
Stakeholder, 63
  characteristics, 167
  feedback, 115
  influence, 167
  management strategies, 169
  position, 169
  power, 168
  predictability of behavior, 169
Stakeholders, 109, 117, 143, 161, 165
  examples, 167
  primary and secondary, 165
  sociopolitical, 166
Stakes, 167
  subjective well-being, 31
Suppliers, 167
Sustainability, 54
Symbols, 39

T
Target market, 123
Technologies
  big brother, 73
  seductive, 73
Throughput time, 157
Transaction governance, 179
Transactions, 158
Transdisciplinarity, 101
Transformation, 90
  age of, 6
  economy, 9
Transformational
  innovation, 130
Transformational innovation, 62, 66
Transformations, 27
Trust, 146, 184

U
Uncertainty, 77, 104
Users, 109

V
Value, 21
  absolute, 38
  balancing, 182
  constellation, 146
  constellations, 17

V (*cont.*)
creation, 10, 55, 63
economic, 79
exchange, 25, 63, 158
for money, 29
for society, 18
for the user, 13
for the ecosystem, 17
for the organization, 15
historical, 41
holistic, 97
individual's orientation, 31
intangible, 131
integral view on, 19
labor, 26
network, 77
networks, 17
objective, 22
of biodiversity, 47
of nature, 48
proposition, 62, 66, 75, 82, 87, 92, 97, 110, 117, 131, 159
sentimental, 41
societal, 79
subjective, 22
surplus, 26
symbolic, 41
systems, 38
transcendental, 38
use, 24
web, 77
Value flow, 173
Value flow model, 154
completeness, 175
Value formula, 137
Value framework, 55, 91, 97
levels, 55
perspectives, 55
value concepts, 56
Values, 21, 39
core, 36
earth-centered, 47
human, 31
motivational, 32
Venturing
corporate, 80
model, 80
process , 125
process model, 127
project, 125
Viability, 104
Virtuous cycles, 78

W
Wealth, 27, 30
Well-being, 37
Wicked problem, 102
Worthiness
perceived, 25

MIX
Papier aus verantwortungsvollen Quellen
Paper from responsible sources
FSC® C105338

FSC
www.fsc.org

If you have any concerns about our products,
you can contact us on
ProductSafety@springernature.com

In case Publisher is established outside the EU,
the EU authorized representative is:
Springer Nature Customer Service Center GmbH
Europaplatz 3, 69115 Heidelberg, Germany

Printed by Libri Plureos GmbH
in Hamburg, Germany